SUSTAINABLE RURAL

Perspectives on Rural Policy and Planning

Series Editors:
Andrew Gilg
University of Exeter, UK
Professor Keith Hoggart
Kings College, London, UK
Professor Henry Buller
University of Exeter, UK
Professor Owen Furuseth
University of North Carolina, USA
Professor Mark Lapping
University of South Maine, USA

Sustainable Rural Systems
Sustainable Agriculture and Rural Communities

Edited by
GUY M. ROBINSON
Kingston University, London, UK

Routledge
Taylor & Francis Group

LONDON AND NEW YORK

First published 2008 by Ashgate Publishing

2 Park Square, Milton Park, Abingdon, Oxon OX14 4RN
711 Third Avenue, New York, NY 10017, USA

Routledge is an imprint of the Taylor & Francis Group, an informa business

First issued in paperback 2016

British Library Cataloguing in Publication Data
Sustainable rural systems : sustainable agriculture and
    rural communities. - (Perspectives on rural policy and
    planning)
    1. Rural development - Congresses 2. Sustainable
    agriculture - Congresses
    I. Robinson, G.M. (Guy M.) II. International Geographical
    Union. Commission on the Sustainability of Rural Systems
    III. International Geographical Union. Congress (30th :
    2004 : Glasgow)
    307.1'412

Library of Congress Cataloging-in-Publication Data
Sustainable rural systems : sustainable agriculture and rural communities / edited by Guy M. Robinson.
    p. cm. -- (Perspectives on rural policy and planning)
    Includes index.
    ISBN: 978-0-7546-4715-7
  1. Sustainable agriculture 2. Sociology, Rural. I. Robinson, G.M. (Guy M.)

    S494.5.S86S897 2008
    338.1--dc22

                                                            2007035369

ISBN 978-0-7546-4715-7 (hbk)
ISBN 978-1-138-25758-0 (pbk)

Contents

PART 3: SUSTAINABLE RURAL COMMUNITIES

List of Figures

List of Tables

List of Contributors

Mary Cawley is a Senior Lecturer in Geography at the National University of Ireland, Galway. Her research relates principally to rural social change and development, notably tourism and population issues.

Rosie Cox is Lecturer in London Studies at Birkbeck, University of London. She has a long-standing research interest in the organisation of paid domestic employment and is author of *The Servant Problem: Paid Domestic Work in a Global Economy* (I.B. Tauris, 2006). She has recently been working with colleagues on a project examining producer/consumer relationships in 'Alternative Food Networks'.

Elizabeth Dowler is Reader in Food and Social Policy in the Department of Sociology, University of Warwick. She works on poverty, food, nutrition and public health; evaluating policy and local initiatives; people's perspectives. Publications include Dowler and Spencer (eds) *Challenging Health Inequalities* (Policy Press, forthcoming); chapters in Dora (ed.), *Health, Hazards and Public Debate (BSE/CJD lessons)* (WHO, 2006); Dowler & Jones Finer (eds) *The Welfare of Food* (Blackwell, 2003); and Mosley & Dowler (eds) *Poverty and Social Exclusion in North and South* (Routledge, 2003).

Nick Evans is a Principal Lecturer in Geography and Director of the Centre for Rural Research at the University of Worcester. He has a long-standing interest in the geography of agriculture. Apart from rare breeds of livestock, he has published articles on farm-based tourism, women in farming and environmental policy, together with more conceptual pieces on 'post-productivism' and 'agri-cultural' geography. He is currently engaged on work to monitor protected landscapes.

Desmond A. Gillmor is Emeritus Associate Professor and Fellow in Trinity College, University of Dublin. His research has been mainly on agricultural geography, tourism and rural development and conservation.

Isabel Griffiths has been involved with the organic movement for years, working initially as an inspector, prior to carrying out this research whilst at Kingston University London. She is currently Production Standards Coordinator at the Soil Association.

Frances Harris is an environmental geographer at Kingston University London and has carried out research on the sustainability or farming systems in Africa and the UK. Research has included work on nutrient balances of farming systems, promoting farmer participatory and interdisciplinary research strategies, and organic farming. She is the editor of *Global Environmental Systems* (Wiley, 2004).

Lewis Holloway is Lecturer in Human Geography in the Department of Geography, University of Hull, UK. He has research interests in food, farming and the countryside, focusing particularly on 'alternative' modes of food provisioning and rural lifestyles, and on the implications of 'high-tech' interventions in livestock farming for human-nonhuman relations in agriculture.

Moya Kneafsey is a Senior Research Fellow in Human Geography at Coventry University. She has published research on 'alternative' food networks and rural tourism, especially in remote and peripheral areas.

Lois Mansfield is a Principal Lecturer in Environmental Management at the Newton Rigg Campus of the University of Cumbria. She has worked in the land-based education sector since 1994 where vocational excellence sits comfortably alongside theoretical understanding. Her interests lie in upland agriculture and countryside management.

Bruce D. Pearce is Deputy Research Director of the Elm Farm Organic Research Centre (EFORC). He has a degree in Biological Sciences and a PhD in Plant Physiology. He joined EFORC (Berkshire, UK) in 1999 as Head of Research. The Centre undertakes a wide range of research into organic and sustainable food production. Prior to joining EFORC he was part of MAFF's Chief Scientists Group, Horticulture Unit and worked for the Consumers Association's *Gardening Which?* magazine as a researcher. He has sat on a number of government advisory panels including the UK Government's GM science Review Panel and as part of an expert group for the Cabinet Office Strategy Unit's study on the costs and benefits of GM crops.

Clive Potter is Reader in Rural Policy in the Centre for Environmental Policy, Imperial College London. His research interests focus around EU, North American and Australian rural policy reform, agricultural restructuring and its environmental consequences. His recent publications have been on the theorisation of agricultural multi-functionality and global trade liberalisation and include papers in the *Journal of Rural Studies* (with Jonathan Burney), *Progress in Human Geography* (with Mark Tilzey), and the *Geographical Journal*.

Guy M. Robinson is the Inaugural Professor of Geography at Kingston University, London. He is the author of *Conflict and Change in the Countryside* (Wiley, 1994) and *Geographies of Agriculture* (Pearson, 2004). He has worked on various aspects of rural development and change in the UK, Europe, Australasia, and North and Central America. His most recent publications have been on agri-environmental schemes in Canada, wine tourism in Australia, and rural change in post-Dayton Accords Bosnia-Hercegovina. He is the UK representative on the Committee of the IGU Commission on Sustainable Rural Systems, and is the Editor of *Land Use Policy*.

Christopher Short is Senior Research Fellow in the Countryside & Community Research Unit and Senior Lecturer within the Department of Natural and Social Sciences at the University of Gloucestershire. He has considerable experience of rural research, much of it associated with the multi-functional use of land. Recent projects include a national review of Local Access Forums (2005), the economic evaluation of the Countryside Stewardship Scheme (CSS) (2000), evaluation of the Upland Experiment (2004) and the monitoring and evaluation of both CSS and Environmentally Sensitive Areas (2002 and 2004). He has published widely on issues concerning common land and assisted the government in the formulation of the Commons Act 2006 as well as being convenor for the National Seminar on Common Land and Village Greens, which brings together a wide range of stakeholders to discuss policy and management issues with a view to improving the management of commons and village greens.

Mark Tilzey is Senior Research Fellow at the New Economics Foundation. His research interests centre on the political economy of rural change, with particular reference to the EU, Australia and the USA, and changing configurations of society-nature relations in rural space. Recent publications have focused on the analysis of agricultural multi-functionality and neo-liberalism and include papers/chapters in the *International Journal of Sociology of Agriculture and Food*, *Progress in Human Geography* (with Clive Potter) and *International Perspectives on Rural Governance* (with Clive Potter) edited by Cheshire, Higgins and Lawrence (Routledge).

Helena Tuomainen is a Research Fellow in the Department of Sociology at the University of Warwick. Her research interests include the sociology and anthropology of food, diet and culture, and the sociology of health and illness. Her recent research examines the relationship between migration, food-ways, ethnic identities and gender through a case study of Ghanaians in London. She specialises in qualitative, ethnographic research and has previously worked on a number of health-related projects.

Laura Venn works as a cultural research analyst at the West Midlands Regional Observatory. With a background in qualitative social research she is responsible for collating and coordinating research and intelligence in relation to the scope and significance of the cultural and creative sectors in the West Midlands and for feeding this into evidence-based regional policy.

Richard Yarwood is a Reader in Geography at the University of Plymouth. He specialises in rural geography and, with Nick Evans, has written widely about livestock geographies. His work in this area has focused on rare and local breeds and has been funded by the Nuffield Foundation, RBST and the Countryside Council for Wales. Other areas of interest include policing, the voluntary sector, housing and service provision in rural areas.

Preface

The majority of chapters in this book were originally presented at sessions on 'Sustainable Rural Systems' as part of the International Geographical Congress held in Glasgow in 2004. The sessions were organised on behalf of two of the commissions of the International Geographical Union (IGU), namely the Commission on the Sustainability of Rural Systems (CO4.33), first established as a Study Group in 1993, and the Commission on Land Use Cover Change (CO4.24). I am grateful to the Committees of both Commissions for their assistance in compiling the programme for the conference sessions, and especially to the former Convenor of the Commission on Land Use Cover Change, Professor Sandy Mather. Sadly Sandy died in late 2006, having been too ill to attend the IGU's Regional Conference in Brisbane a few months earlier. His lasting influence on work on land use policy and decision-making will be long remembered and his support for the work of both Commissions will be much missed.

I am grateful for the assistance provided by all the authors of the individual chapters herein, especially Nick Evans, Richard Yarwood and Bruce Pearce who supplied material not presented in Glasgow. I also wish to acknowledge assistance from Claire Ivision for her cartographic inputs.

PART 1
Introduction

Chapter 1

Sustainable Rural Systems: An Introduction

Guy M. Robinson

Introduction

Since it was popularised by the 1987 World Commission on Environment and Development (WCED), generally known as the Brundtland Report, the term 'sustainable development' has become one of the most widely used by governments and international organisations. The WCED referred to sustainable development as "development that meets the needs of the present, without compromising the ability of future generations to meet their own needs" (WCED, 1987). However, both this definition and the very concept itself have been much criticised (Carley and Christie, 2000; Sachs, 1999). At the heart of the critique are the inherent contradictions between 'sustainable' and 'development': can conservationist ideals contained within environmental notions of 'sustainable' be married to conceptions of development and economic growth? In some circles this has led to greater emphasis upon the notion of 'sustainability', divorcing it from the more problematic 'development' (Callicott and Mumford, 1997).

Post-1987, definitions of sustainable development in terms of moral obligations to guarantee the quality of life for future generations have clarified the meaning of the term "to encompass the full array of social, economic and environmental relations" (Bowler *et al.*, 2002: 5). However, the relative priorities assigned to each of these three intersecting domains have been strongly contested (Adams, 1995). Moreover, the term has been widely used in cavalier fashion to justify actions that stray far from the original sentiments of the WCED. Most notably governments and some non-governmental organisations (NGOs) have used it to refer to long-term maintenance of economic growth rather than to environmental or social dimensions of sustainable development, thereby severely devaluing the term's currency (Trzyna, 1995). Thus, "sustainability is used at once to legitimate calls for unbridled economic growth, industrial expansion, globalisation, the protection of biodiversity, maintenance of ecosystems, social justice, peace and the elimination of poverty" (Bowler *et al.*, 2002: 5). At the very least this suggests a paradox in that sustainable development is deemed to support both the maintenance of the status quo and radical change. In part, this contradiction is linked to the many and different strands of philosophical thinking on society-nature relations (Robinson, 2002a). When moving from general sentiments to a specific context, more exacting questions must be answered:

"what exactly is being sustained, at what scale, by and for whom, and using what institutional mechanisms?" (Sneddon, 2000: 525).

Nevertheless the term continues to be applied in various contexts as a multi-dimensional one encompassing multi-faceted challenges. These were addressed on a global stage in the United Nations World Sustainable Development Summit held in Johannesburg in 2002. Here it was acknowledged that: "Sustainable development calls for improving the quality of life for all of the world's people without increasing the use of our natural resources beyond the Earth's carrying capacity. It may require different actions in every region of the world, but the efforts to build a truly sustainable way of life require the integration of action in three key areas" (UN, 2002: 6).

These areas were identified as:

- Economic growth and equity. Global systems demand an integrated approach to foster responsible long-term growth while leaving no-one behind;
- Conserving natural resources and the environment. There is a need to introduce economically viable solutions to reduce resource consumption, stop pollution and conserve natural habitats;
- Social development. There needs to be respect for the rich fabric of cultural and social diversity and the rights of workers. All members of society must be empowered to play a role in determining their fortunes.

The debates at the Summit built on related deliberations of world leaders at the United Nations Millennium Summit of 2000 and the previous Earth Summit in Rio de Janeiro in 1992 to develop concrete proposals for countries to re-examine their consumption and production patterns, commit to responsible, environmentally sound economic growth, and work together purposively to expand cross-border cooperation and share expertise, technology and resources. However, whilst it is possible to identify 'advances' in implementing legislation that fosters environmental (as opposed to economic) sustainability, the world's growing demand for food, water, shelter, sanitation, energy, health services and economic security have widened global inequalities and thereby restricted any concerted moves towards greater sustainability. This can certainly be seen with respect to agriculture where industrial-style production is dominant, especially in the Developed World, and attempts to promote sustainable agriculture have not been implemented widely (Robinson, 2002b). Similarly, with respect to human settlements, attempts to foster reduced resource consumption and the creation of 'sustainable lifestyles' have generally foundered despite the emergence of various international agreements to deliver environmental benefits, perhaps most notably the Kyoto Protocol, an agreement made under the United Nations Framework Convention on Climate Change. The 160+ countries that have ratified this protocol commit to reduce their emissions of carbon dioxide and five other greenhouse gases, or engage in emissions trading if they maintain or increase emissions of these gases.

Geographers, with their concerns for both the physical resource-base and human dimensions of sustainability, have played a significant role in academic discourse on sustainable development. This can be seen throughout the 1990s in various conferences in which themes relating to sustainable development were a frequent occurrence. Within Geography's principal worldwide organisation, the International Geographical Union (IGU), one illustration of this growing engagement with debates on sustainable development can be seen in the establishment of the Commission on the Sustainability of Rural Systems (CO4.33) as a Study Group in 1993. The Commission subsequently developed a programme of annual international and regional conferences, primarily involving researchers from rural geography, but also attracting academics and practitioners from planning, resource management, politics and information technology. At the IGU's 30th Congress, held in Glasgow in 2004, the Commission's sessions included a range of papers relating to sustainable agriculture and sustainable rural communities. Given the location of the Congress, it is not surprising that these papers had a bias towards issues pertinent to the United Kingdom (UK) and with a dominant representation from UK-based geographers. This book brings together some of these papers from the Commission's sessions in Glasgow, re-creating the key themes covered and the debates on sustainable rural systems, with a particular British dimension.

The starting point, by Mark Tilzey and Clive Potter (Chapter 2), is an analysis of the background macro-scale processes of socio-economic and political change in which are embedded the specific studies represented by the book's individual chapters. They outline key features in the agri-food system as discussed in current analyses of the sustainability of agriculture in the Developed World. Hence they refer to the transitions from productivist to post-productivist agriculture, from Fordism to post-Fordism, neo-liberal economic management, and both systemic and non-systemic moves towards greater sustainability. This broad and rapidly changing tapestry is given more concrete expression by specific focus on how post-productivism (essentially representing moves away from an overwhelming commitment to increased farm output) is manifest in the European Union (EU), the United States and Australia.

They describe post-productivism in the EU as conforming to an 'embedded neo-liberal' mode of governance. Essentially this means that the EU's Common Agricultural Policy (CAP) has been partially reformed to inject environmental and rural development objectives within ongoing modifications to the previously strongly protectionist approach to European farming. So the CAP is becoming progressively more market oriented through reductions in price support and direct payments to farmers. This can be seen in the Rural Development Regulation in which there are priorities for assisting adaptation to more market-oriented agriculture and dealing with risk in competitive markets. There are distinctive regional consequences to these reforms, with strong contrasts emerging between those areas dominated by productivist industrial-style farming, such as East Anglia, the Paris Basin, the Low Countries and the Po Valley, and marginal and upland zones where farmers are more likely to be focused on the production of environmental goods and supplying niche markets. Various environmental payments are unlikely to compensate for the declining incomes of the largely small and medium-sized farmers in these marginal/upland areas and so their continuing demise seems likely. Clearly this is not a socially or economically sustainable situation and there must also be question-marks over the long-term environmental benefits that can be realised without a vibrant farming community. Tilzey and Potter conclude that different types of policy intervention will be required to move towards greater sustainability, but that these "are considered increasingly illegitimate under neo-liberal norms".

An even greater adherence to neo-liberalism is reported for Australia, with the overwhelming dominance of productivism orientated towards exports and externalisation of environmental and social costs. Output-based supports and protectionist policies have been abandoned from the 1970s, with a Rural Adjustment Scheme (RAS) from 1977–88 helping farmers adjust to free-market determined prices. Neo-liberalism as applied to the Australian economy since 1983 (Robinson *et al.*, 2000: 242–5) has encouraged increased farm output, but at the expense of environmental quality in many cases. Indeed, the long-term ability of the extensive rangelands to support livestock production has been questioned (Hamilton, 2001). However, mitigating measures have been quite restricted. The introduction of the Landcare scheme in 1989 attracted international attention and has emphasised the need for local communities to identify key environmental problems and to derive solutions drawing largely on local resources (Lockie, 1998). However, Tilzey and Potter argue that core problems of land degradation are not being addressed.

They view the United States as representing a hybrid between the EU and Australia, sharing with the former a strong differentiation between the lobbies representing small/ medium-sized farmers and larger operators, but with an overwhelming dominance of productivism. Its agricultural policies have consistently endeavoured to maintain family farming to produce food for the domestic market whilst allowing market forces to develop scale economies and production for export. Family-based enterprises still dominate production and represent a powerful lobby that government cannot ignore. This has helped to retain certain forms of protectionism and few attempts to reign in the productivist system. For example, federal policy has largely ignored calls to maintain biodiversity and landscape quality in conjunction with agricultural activity as these are deemed to exist outside or in opposition to agricultural practice. The dominant form of expenditure on farm-related environmental programmes has been in the form of promoting the retirement of land from farming or, under the 1996 and 2002 Farm Bills, mitigating the impacts of pollution from productivist agriculture.

Nevertheless, the Conservation Security Program (CSP), introduced in the 2002 Farm Bill, may be significant as it links conservation directly to on-farm production in a similar fashion to agri-environment policy in the EU. In support of this policy there has emerged a sustainable Agriculture Coalition, prominent outside what Tilzey and Potter term "the zones of mass (productivist) food commodity production", notably in New England where there is some intersection between post-productivist farming and consumers concerned with food quality. At present the CSP has a strict budgetary cap, which has restricted its potential to counter continued reliance on conventional productivist farm support on the small and medium-size farms where production of environmental goods might be most attractive. Its current relatively marginal status within the overall ongoing support for productivist agriculture is a reflection of the continuing dominance of neo-liberalism in the US. This dominance can be seen in continuing support for measures designed to increase world market access for US produce. However, this is contradicted in some continuing domestic market distortion through interventions designed to support certain groups of producers experiencing problems associated with declining global agricultural commodity prices.

Tilzey and Potter conclude that in the EU, Australia and the United States productivism remains dominant. There may be pockets of post-productivism, but attempts to engender greater sustainability and post-productivism have been subordinate to, or defined by, more dominant concerns that prevent synergies from developing between the economic, social and environmental elements of sustainability. Significantly, they also conclude that post-productivism therefore cannot be conflated with post-Fordism. The latter represents new forms of regulation, a new techno-economic paradigm and new forms of production. Some of these changes may have permitted or encouraged the emergence of post-productivist agriculture, but not at the expense of destroying the dominance of productivism. Therefore to understand the emergence of (minority) discourses privileging sustainability and policies promoting on-farm environmental actions, the particular contexts and political circumstances in which these occur need to be analysed. In effect, the succeeding chapters of this book provide such analysis, drawing upon particular geographical contexts to analyse specific developments within post-productivism and the emergence of more sustainable agricultural and rural systems.

Sustainable Agriculture

Despite the promotion of more extensive forms of farming in some parts of the Developed World, the production of 'environmental goods' by farmers and a greater concern over the quality of food being produced, Bowler (2002b: 180) contends that "even taken together, the emergent features... do not constitute the basis for sustainable farming systems." He contends that a truly sustainable agriculture must represent a clear alternative to the industrial model as part of a transformation of both the farm economy and the society in which it is embedded. However, there are many different models that have been proposed as representing an 'alternative', embracing a range of philosophies on sustainable farming, including organic, ecological, biodynamic, low-input, perma-culture, biological, resource-conserving and regenerative systems. Determining which of these is sustainable and how they differ from other alternatives depends on exactly how 'sustainable agriculture' itself is conceptualised.

With respect to agriculture in the Developed World, it has been the industrial agri-food system that has been dominant post-1945. This system is efficient and effective in economic terms, but in social and ecological terms it is not sustainable (Troughton, 2002). Any analysis of the system highlights various negative feedbacks that are inimical to environmental and social dimensions of sustainability and hence various 'alternative' systems, such as organic farming, are proposed to overcome the unsustainable nature of 'conventional' food production and consumption systems. 'Unsustainability' has been associated with the key elements of the industrial or productivist agricultural system: intensification, concentration and specialisation (Bowler, 2002a), all of which can be linked to some degree to national and supra-national agricultural policies. These have subsidised intensification, encouraged investment in new technology, funded advisory services which have promoted diffusion of new farming technology and developed the production of this new technology.

There has been substantial research on the limits to the sustainable development of productivist agriculture. This has recorded a range of environmental disbenefits, including diminished biodiversity, removal of natural and semi-natural habitats, rising soil erosion and salinity, reduced water tables, pollution of water courses and growing reliance on an excessively narrow range of crops and livestock. The threats posed to the environment vary in magnitude and type between productivist farming systems. Similarly, there has been great variability in the extent to which the industrial productivist model has delivered economic sustainability. Indeed, despite the substantial support mechanisms directed towards farmers and the farming sector in North America, the EU and Japan, profit margins have been squeezed. This has been most apparent in the livestock sector and on small and medium-sized family-run holdings, where the movement of people off the land has been substantial for at least four decades. The gap between farm and non-farm incomes has risen as profits in the agri-food sector have been concentrated increasingly in the upstream (farm supply industry) and downstream (wholesale and retail) sectors.

The flight from the land has accompanied other trends fostering rural depopulation and so reducing the social sustainability of both farming and rural communities. Therefore it is possible to identify a number of negative feedback loops within the productivist system that have contributed to reduced agricultural (and rural community) sustainability. Policy-makers and farmers alike have responded to this by addressing various 'negativities', generally though focusing upon particular components of unsustainability rather than offering holistic solutions. Examples include attempts to inject environmentally beneficial methods of farming, the promotion of farm diversification, a growing concern for food quality and 'technical fix' methods to raise output. Clearly some of these reinforce the productivist model whilst others offer some form of 'alternative' that has been viewed in some circles as the emergent features of sustainable agriculture.

A general conclusion is that moves towards sustainable agriculture need to develop a focus beyond a disintegrated framework of local developments (Bowler, 2002b). This may entail new forms of social contract between food producers, food retailers, food consumers and the state. Necessarily, at the heart of this 'contract' is a reconnection of consumers to the sources of their food supply after the dislocation associated with the productivist agri-food system. However, only with strong regulation will sustainable agriculture emerge except on a highly adventitious and spatially fragmented basis. This notion of the importance of regulation is part of recent attempts to enhance sustainability through 'ecological modernisation', in which environmental regulation and economic growth are regarded as mutually beneficial if the demands of each are carefully balanced (Giddens, 1998). It is not clear, though, how maintenance of the primacy of productivist imperatives can be balanced successfully with concerns about public health, the environment and farm welfare (in the process of ecological modernisation) to deliver sustainable agriculture.

Of course this begs the question of 'what is sustainable agriculture?' In referring to the difficulty in defining the term, O'Riordan and Cobb (1996) point out that not only do we have limited knowledge about the ecological nature of a 'sustainable farming system' in terms of calculating the full profit and loss account of natural resources but there are also marked divergences of opinion between farmers and other interested parties. When questioned about sustainability, farmers tend to focus on economic factors rather than environmental ones and do not recognise the need to reduce purchased inputs or energy use. In contrast, academics stress the core ecological values of sustainable agriculture (Dunlap *et al.*, 1992). Key problems in defining sustainable agriculture are:

- The difficulty in determining the nature of sustainable soil characteristics.
- The contribution of a sustainable agriculture to biodiversity.
- The lack of sufficient inter-disciplinary research on this topic (Cobb et al., 1999).

As good a definition as any is provided by Francis and Younghusband (1990: 230):

> Sustainable agriculture is a philosophy based on human goals and on understanding the long-term impact of our activities on the environment and on other species. Use of this philosophy guides our application of prior experience and the latest scientific advances to create integrated, resource-conserving, equitable farming systems. These systems reduce environmental degradation, maintain agricultural productivity, promote economic viability in both the short- and long-term, and maintain stable rural communities and quality of life.

This can be compared with the conditions to be satisfied if agricultural systems are to be satisfied, as shown in Tables 1.1 and 1.2. Compared with the current commercial, industrialised model of farming there are clear differences, with sustainable agriculture emphasising limited inputs, use of specific soil preserving practices (e.g. no-till systems) and management perspectives based on ecological and social considerations (e.g. biodynamics and permaculture) (Bowler, 2002a; 2002b).

Research in southern Africa suggests that sustainability relies on a judicious combination of resource-conserving technologies accompanied by concerted attempts to sustain local institutions and to create an 'enabling' environment whereby smallholders have secure access to land, research and development on sustainability, and also reliable markets (Charlton, 1987; Whiteside, 1998). Most of the work on sustainable agriculture focuses on particular environmental, economic or social criteria, though elements of all three are needed in tandem for the creation of a self-sustaining system, as recognised in terms of three key components according to Brklacich et al (1997):

- Environmental sustainability or the capacity of an agricultural system to be reproduced in the future without unacceptable pollution, depletion or physical destruction of its natural resources and natural or semi-natural habitats.
- Socio-economic sustainability or the capacity of an agricultural system to provide an acceptable economic return to those employed in the productive system.
- Productive sustainability or the capacity of an agricultural system to supply sufficient food to support the non-farm population.

Table 1.1   Conditions to be satisfied if agricultural systems are to be sustainable

---

- Soil resources must not be degraded in quality through loss of soil structure or the build-up of toxic elements, nor must the depth of topsoil be reduced significantly through erosion, thereby reducing water-holding capacity.
- Available water resources must be managed so that crop needs are satisfied, and excessive water must be removed through drainage or otherwise kept from inundating fields.
- Biological and ecological integrity of the system must be preserved through management of plant and animal genetic resources, crop pests, nutrient cycles and animal health. Development of resistance to pesticides must be avoided.
- The systems must be economically viable, returning to producers an acceptable profit.
- Social expectations and cultural norms must be satisfied, as well as the needs of the population with respect to food and fibre production.

Source: Benbrook, 1990.

Table 1.2   Differences between sustainable and industrialised agriculture

---

Sustainable agriculture implies

- Less specialised farming. This often requires mixed crop and livestock farming for reduced dependence upon purchased inputs.
- Off-farm inputs should not be subsidised and that products contributing adverse environmental impacts should not receive government price support.
- Farm-level decision-making should consider disadvantageous off-farm impacts of farm-based production, e.g. contamination of groundwater, removal of valued landscape features.
- Different types of management structure, e.g. family farms as opposed to corporate 'factory' farms.

Source: Dearing, 1992.

There is also a spatial dimension to this conceptualisation of sustainable agriculture, with Troughton (1993; 1997: 280) recognising a five-fold nested spatial hierarchy of sustainability (see also Lowrance, 1990), as shown in Table 1.3.

Similar conceptualisations have been less hierarchical, notably the five-fold categorisation employed by Buttel et al (1990) in selecting criteria to be measured or assessed for their degree of sustainability:

- The nature of the productive process.
- The economic and social organisation of food production.

Table 1.3   A spatial hierarchy of sustainability

- Agronomic sustainability: the ability of the land to maintain productivity of food and fibre output for the foreseeable future (field-level).
- Micro-economic sustainability: the ability of farms to remain economically viable and as the basic economic and social production unit (farm-level).
- Social sustainability: the ability of rural communities to retain their demographic and socio-economic functions on a relatively independent basis (community-level).
- Macro-economic sustainability: the ability of national production systems to supply domestic markets and to compete in foreign markets (country-level).
- Ecological sustainability: the ability of life support systems to maintain the quality of the environment while contributing to other sustainability objectives (global level).

Source: Troughton (1997).

- The role of scientific research and extension activities.
- The use and management of labour.
- The organisation of marketing and distribution activities.

In addressing the key issues relating to the development of more sustainable agricultural systems, it is widely agreed that there is an urgent need for a greater input of ecological and environmental information in agricultural policy. At present much research on sustainable agriculture focuses on individual plots or fields or else it is performed at farm level. Yet, most of the environmental issues associated with agriculture are manifested at larger scales, and there is a need to consider measurable changes in water quality, soil quality, biodiversity and other environmental indicators at these scales (Allen *et al.*, 1991). In addition to this need to understand the operation of sustainability at different spatial scales, we also need to have more knowledge of what sustainability means to farmers. The following is a series of questions proposed as part of developing a research agenda for sustainable agricultural systems in the late 1990s (Robinson and Ghaffar, 1997). A decade later they are still equally pertinent:

- How can we measure sustainability at a variety of scales: farm-level, regional, national, supra-national?
- How can notions of 'sustainability' be built into agricultural policy other than by simply promoting environmentally friendly farming or landscape restoration within the overall context of an essentially economic agricultural policy?
- To what extent do existing agri-environmental measures contribute to achieving specified objectives (such as the maintenance or enhancement of biodiversity and the conservation of cultural heritage)?
- Can the implementation of environmental strategies in peripheral areas assist in supporting threatened traditional forms of input-extensive management?

The initial evidence from the EU's Environmentally Sensitive Areas (ESAs) scheme and Tir Cymen in Wales seems quite positive.

- To what extent do selected measures contribute towards or reinforce a polarisation of the landscape into agriculturally productive areas and areas primarily producing environmental goods?
- Do certain measures assist a reduction of agricultural surpluses by promoting general extensification?
- What is the socio-economic impact of agri-environmental measures on farms, especially in peripheral areas?
- To what extent can such measures become an integral part of the survival strategy of individual farmers?
- To what extent do they contribute to diversification and sustainability?

Five of the chapters in this book consider various aspects of sustainable agriculture relating to these questions through discussion of key related topics, such as alternative food networks, organic farming, the introduction of genetically-modified crops, the role of farm livestock in sustainable agriculture, and multifunctional land use.

Alternative Food Networks

One potential avenue for the development of greater rural sustainability is that of alternative food networks (AFNs). These have been defined in various ways, but essentially as the direct opposite of 'conventional' agri-food systems. This means that AFNs are characterised by production, processing, distribution and consumption of food within a particular region. This is therefore a 'short' food chain as opposed to the globally-sourced, food miles-generating, long-distance supply chains that typify many products sold in the major supermarkets.

Growth of AFNs, apparent from the late 1980s, has occurred as a direct result of local ('bottom up') initiatives aimed at generating closer links between food producers and consumers in the region or locality of production. So at one level there are numerous such initiatives that might be labelled as AFNs, but without agreement on the extent to which they represent a true 'alternative' to the conventional system.

Goodman (2003) recognises differences between an American model of AFNs and a European model. The former represents a more radical challenge to the conventional food supply system in an explicitly politicised activism. This links AFNs directly to political challenges to the globalised capitalist system by offering a well-defined alternative. In Europe and the UK that alternative is somewhat 'fuzzy' as it tends to exist more at the margins of the conventional system and, so far, has not offered a concerted challenge to that system. Indeed in continental Europe and the United States, retailing of organic produce has generally been associated with small-scale, 'local' outlets rather than large supermarkets, and so there has been a clear link between organic farming and ethical shopping that supports particular types of production and retailing (Young and Welford, 2002). In contrast, in the UK, supermarkets have played a major role in the organic sector, staring with sales of produce labelled as 'organic' by Safeway in 1981.

The growth of AFNs reflects key changes in consumer behaviour that emerged during the 1990s (Dawson and Burt, 1998: 163). First, there has been a search for individualism in lifestyle through products and services purchased. This represents a counter to the tendency towards a uniform pattern of mass demand that is typified in the presence of the same chain of restaurants in all major urban centres around the world. Secondly, there are higher expectations of the quality of products, services and shopping environments. A degree of non-price responsiveness to advertising and promotional methods can be added to this and, finally, less direct price comparison, with an increase in product range and differentiation. In part these developments have contributed to growing consumer resistance and reaction to processed, manufactured and fast food. Hence it is possible to present several dichotomies that are combining to produce the current trajectory of the wider agri-food system: productivism versus post-productivism, industrial versus alternative, global versus local. The last of these three contrasts is explained partially by the growing concern of the general public for the quality of the food they are consuming, and a willingness of some to pay a price premium for some guarantee of quality (Evans *et al.*, 2002: 317).

This focus on quality can be explained with reference to several inter-related factors. First, foods that can be identified in some fashion as being 'environmentally friendly' can allay consumers' worries about 'scares' over harmful diseases associated with agriculture and farming's negative impacts on the environment. In addition, the purchase of 'quality' foods enables certain groups of consumers to differentiate themselves. In this way they can incorporate the act of purchasing food into a statement of their own identity. This represents an enhancement of their 'cultural capital' as distinguished in the phrase 'we are what we eat' (Bell and Valentine, 1997). For producers, the development of 'quality' foods has provided a marketing opportunity by adding value to the product and developing supply-chain differentiation from their competitors (Nygard and Storstad, 1998). As part of this, major food retailers have encouraged quality assurance schemes as one element in new approaches to the management of supply chains so that retailers can ensure their market share and maintain a competitive advantage. Partly because of the various 'food scares' in recent years, the focus upon quality within supply-chain management has become part of a crucial insurance policy for the major food retailers.

This has coincided with the emergence of a distinctive ethical consumerism in which a sub-set of consumers has expressed concerns over the environmental and social impacts of their consumption practices. These concerns have taken the form of personal implementation of consumption patterns that some commentators have termed 'sustainable' (Murdoch and Miele, 1999). This includes various forms of 'green' and 'ethical' behaviour, e.g. buying phosphate-free detergents, goods with minimal packaging, goods without pesticide additives, and 'fair-trade' products.

Chapter 3 by Cox *et al* investigates how AFNs represent a particular construction of sustainability by virtue of 'reconnecting' producers to their local community and consumers to local producers. Their case studies are five schemes in various parts of the UK, which all involve direct contact between producers and consumers. They include provision of vegeboxes, farm shops and sales at farmers' markets. Interviews with sets of consumers and producers reveal a variety of environmental, economic and social motives underlying 'sustainable' consumption. However, a common theme is the conscious examination of the broad consequences of particular consumption habits. This can involve considerations such as a desire for a more healthy diet, the greater satisfaction gained from direct interaction with producers of food (as opposed to the impersonal quality of purchasers from a supermarket), and a concern for purchases to be more 'environmentally friendly'. Cox *et al* regard these concerns as forming essential dimensions of what they call 'a conscious consumer'. There are some potentially negative consequences of this consciousness for the consumer, namely: seasonal restrictions to certain purchases, greater time and effort for a purchasing process involving farmers' markets and farm shops, greater organisation than compared with supermarket purchases, and greater costs. Access to schemes involving AFNs is not always readily available to the public as a whole or may be perceived as a substantial barrier when compared with the ease of supermarket shopping.

Sustainability and Farm Livestock

Evans and Yarwood (Chapter 4) argue that most conceptions of agricultural sustainability largely overlook the position of farm livestock. Despite concerns over the impacts of industrial-style farming methods upon livestock, for example through production systems such as intensive broiler units and the rapid transmission of animal diseases, such as foot-and-mouth affecting cloven-hooved animals, the role of farm animals within moves towards sustainable agriculture has been neglected. Hence the wider implications of long-term genetic changes in breeds of livestock from the early nineteenth century through selective breeding programmes have been largely overlooked. Ultimately these changes have led to a loss of breed diversity and therefore, arguably, decreased sustainability.

Evans and Yarwood chart the specific outcomes for livestock arising from the three structural dimensions of productivism (intensification, concentration and specialisation). They examine the outcomes systematically for dairy cattle, beef cattle, rare breed cattle, sheep and pigs, after noting the apparent lack of public concern over the loss of breed diversity, with just 8,000 members of the Rare Breeds Survival Trust (RBST) contrasting to over one million members of the Royal Society for the Protection of Birds (RSPB). Even the membership of Butterfly Conservation is larger than that of the RBST! The chapter notes that more breeds of cattle are within the most endangered categories on the RBST's survival watch-list than any other type of farm animal. However, this list does not necessarily reflect just breeds that are genetically rare, but includes some, such as Devon Cattle (Ruby Red), that have a particular local geographical association and, as a result, are viewed as important in terms of sustainability.

Agri-environment schemes in the UK have neglected livestock generally and rare breeds specifically, but there are more livestock-friendly schemes in other countries, such as the Rural Environment Protection Scheme (REPS) in Ireland, in which the latest version has a measure devoted to conservation of animal genetic resources (rare breeds). It was not until 2006 that a similar concern was expressed within a policy in the UK, in the form of the Environmental Stewardship Higher Level Scheme, with funding to support the keeping of native breeds at risk. In contrast, the UK's

Biodiversity Action Plans (BAPs) ignore farm animals except where identifying overgrazing as a threat to wild flora. So it has tended to be non-policy-related developments that have played a much more significant role in the conservation of rare breeds as part of the commodification of the countryside associated with the growth of rural tourism, quality food products and farm diversification (Yarwood and Evans, 2000). Real consideration of how farm livestock fit into various environmental scenarios has rarely even begun and so there is huge potential for exploring the grazing habits of individual breeds and their environmental outcomes. It is this potential that needs to be exploited if farm livestock are to be adequately incorporated into moves towards greater agricultural sustainability.

Organic Farming

Although it is an oversimplification of current debates about agricultural sustainability, in terms of moves towards a more sustainable agriculture, there are two contrasting approaches: the ecocentric and the technocentric. The former stresses a direct alternative to industrial-style 'modern' farming methods by advocating a low-growth model of development, such as organic and biodynamic farming, with radical implications for changes in consumption patterns, resource allocation and utilisation, and individual lifestyles. Amongst the changes proposed to farming systems are greater diversification of land use, integration of crops and livestock, traditional crop rotations, use of green or organic manures, nutrient recycling, low energy inputs and biological disease control. Various combinations of these 'alternatives' can be seen in different models of sustainability, including organic farming, permaculture, low input-output farming, alternative agriculture, regenerative farming, biodynamic farming and ecological farming.

In contrast there is an instrumentalist, more technocentric view that rejects the ecocentric as being both practically and politically unrealistic. Essentially one set of alternatives to the ecocentric tends to see sustainable agriculture as a contextual process that may act as a stated goal to be used to modify existing agricultural systems rather than a set of specific prescriptions. These 'modifications' have taken the form of various different types of agriculture, including integrated crop management, diversified agriculture, extensified agriculture and conservation agriculture. In general these advocate an extensive, diversified and conservation-oriented system of farming. Some aspects of these systems have been encouraged in policy initiatives such as limits on applications of fertilisers, imposition of minimum standards of pesticide residues in food, restrictions on types and rates of application of agrochemicals, and subsidies to promote environmentally friendly and lower input-output farming systems.

Of the ecocentric approaches the one that has attracted the most public attention has been organic farming, a production system that avoids or very largely excludes the use of synthetic compounded fertilisers, pesticides, growth regulators, and livestock feed additives (Foster and Lampkin, 2000). Organic farming systems rely on crop rotations, crop residues, animal manures, legumes, green manures, off-farm organic wastes and aspects of biological pest control to maintain soil productivity and tilth, to supply plant nutrients, and to control insects, weeds and other pests. Certification schemes are operated to license organic food production and to ensure that there are appropriate safeguards for consumers.

Although in many parts of the Developed World governments have validated certification schemes for organic farming and have provided financial support, especially during the period when production is being converted from a non-organic mode, only a minority of consumers purchase organically produced food. This partly reflects the price premium usually associated with organic produce, but also the limited supply, with less than 5 per cent of farmers in the EU, for example, having organic certification. Farmers may have been deterred from converting to organic production for a variety of reasons. Boulay (2007) reports that farmers in Normandy regard organic farming as an undesirable return to 'old fashioned' methods of farming as used by their grandparents. Other problems include decreased yields, higher costs because of the increased labour input and environmental issues associated with nutrient leaching, volatilisation of livestock gases and soil imbalances. These can be contrasted against the positive environmental benefits associated with organic farming, including increased and diversified populations of insects, wild flowers, mammals and birds, in addition to enhanced soil structure and reduced soil erosion (Arden-Clark and Hodges, 1988). The higher prices for organic produce may compensate for lower outputs per hectare, and its labour intensity may support more farm workers thereby adding to sustainability of the farm population and rural society.

Sales of organic foods have been the most rapidly growing sector of the market in the Developed World in the past decade, though their overall share of the market is small (less than 5 per cent). Moreover, the movement of farmers into organic production has been highly variable both spatially and temporally. For example, nearly half of the certified organic producers in the United States are in California. The largest amounts of organic cropland are in California, Idaho, Minnesota, Montana and North Dakota. However, organic foods represent less than 3 per cent of US food sales. In the EU organic accreditation is underpinned by European Council Regulation 2092/91, providing specific rules for the production, inspection and labelling of organic products, and with implications for farming systems and the environment. After this Regulation was introduced the land area in the EU under organic production more than tripled within five years, with various organic aid schemes promoting conversion to organic production. Such schemes have proved most popular in Austria, Germany, Italy and Sweden, which all had more than 10,000 certified organic producers by 2000 (Foster and Lampkin, 2000). Conversion rates have been slower elsewhere despite increased subsidisation of conversion, e.g. in 1999 a further £24 million was added to the UK's Organic Farming Scheme, to encourage an additional 75,000 ha of new conversion (MAFF, 2000). Between 2000 and 2006 a further £140 million was allotted for this purpose, providing payments for conversion over a five-year period. This recognises the loss of profits that occurs during the conversion process and provides both compensation for this and a measure of support for costs incurred.

For farmers, key attractions of organic production appear to be both financial (the prospect of premium prices commanded by organic produce coupled possibly with reduced production costs) and environmental (the positive environmental benefits of organic production methods that contribute to notions of good stewardship and care for the land). However, the essay by Harris *et al* (Chapter 5) shows that, in the UK, these attractions have been balanced by various negative factors that have contributed to withdrawal of some farmers from this sector. Their research reports on the end of the phase of rapid expansion of organic production in the UK in the early 2000s and the withdrawal from certification by some farmers. It is these farmers who are quitting organic production that are the focus of this research, with an analysis of the background to and the reasons for farmers terminating their organic certification. Data from two of the leading organic certification agencies are used to show the distribution of both new organic producers and those quitting organic farming. Both incomers and outgoers tend to be from the same part of the UK: south-west and western England, where organic farming is dominated by livestock producers who are focusing on organic meat and organic milk. It is not surprising therefore that one of the chief factors influencing the withdrawals from the sector is the lack of a price premium for organic milk. Indeed, UK milk production in general has been in crisis in recent years because of the dramatic fall in profit margins.

Just over half of the UK's dairy farmers (53 per cent) left the industry between 1995 and 2006. During this time the supermarkets' margin on fresh milk increased from 3p per litre to 16p per litre (DEFRA, 2007). Whilst the costs to farmers for fertilizers, fuel and feed rose substantially their share of the money spent on purchases of milk declined; the share taken by processors and the companies that collect, pasteurise and bottle milk remained about the same; and the share taken by retailers dramatically increased (Mercer, 2006). In part, this reflects the loss of power over retail prices that farmers have been able to exert since the break-up of the Milk Marketing Board (MMB) in 1994. The MMB had held a monopoly on collection and selling of milk, but its demise led to a fragmentation of marketing and processing of milk. This enabled processing companies to compete fiercely to obtain contracts with supermarkets who have tended to work with just one or two suppliers. The pressure this has placed on milk prices has been reflected in the pressure placed on dairy farmers as three processors (Arla, Robert Wiseman and Dairy Crest) now account for two-thirds of the milk sold to the public (primarily via supermarkets). Only half of the milk produced on UK farms is sold as fresh milk to drink. The rest goes into manufacturing, competing with milk produced elsewhere in the global commodity markets. Here, plentiful supply from countries with lower production costs and more favourable exchange rates have helped keep down prices obtained by British producers. The farmers received 24.5p (58.3 per cent) of a litre of milk retailing for 42p in 1995 compared with 18.5p (38.9 per cent) for a litre retailing at 47.5p in 2004 (DEFRA, 2007). To remain viable, full-time dairy farmers have had to adopt scale economies as herds of less than 100 cows have effectively ceased to be competitive. Moreover, reforms to the CAP have reduced subsidies to dairy farmers that helped to protect them from the realities of changes in the global market. It has largely been this global market that has determined the price of milk set by the processors (and hence by the supermarkets). For many organic milk producers, the growing inability to find a market for their milk has meant they have been forced to sell their organic milk as 'ordinary' milk, the price of which has effectively fallen significantly as described above. The economics of this situation has been a crucial factor in decisions by some to leave the organic milk sector.

Harris *et al* record four main sets of reasons underlying farmers' decisions to leave organic certification. These are financial, negative experiences of the certification and inspection process, negative experiences of implementing the organic system on the farm, and miscellaneous factors including the impacts of the Foot-and-Mouth epidemic of 2000/1, changed personal circumstances and the distance to certified organic abbatoirs. The prominence of financial factors is closely linked to the fact that the initial decision to convert to organic production was largely a financial one. Farmers were attracted by a generally correct perception that organic produce commanded 'high prices', that there were new and expanding markets for this produce, conversion grants were available and that conversion therefore represented an attractive financial option to intensification of their existing farming system. Indeed, in the harsh economic climate for farmers, this choice between organic and intensification was frequently viewed as the only pair of options available. Farmers were often further encouraged to adopt organic certification by their perception that they were already operating a farming system that would not require much alteration to obtain certification. Coupled with attitudes that were very sympathetic to the environmental connotations of organic production and the positive financial outlook, the decision to apply for organic certification for many seemed quite logical and attractive. However, it appears that for some of those converting, financial outcomes have been much poorer than expected, the conversion process has been more complex and difficult, and organic farming systems have proved harder to operate than expected.

Harris *et al* conclude that a majority of those leaving organic certification appear to be what they term 'pragmatic' organic farmers (after Fairweather, 1999), motivated largely by the price premiums on organic food. This group can be contrasted against the 'committed' organic farmers, motivated by the organic philosophy and ideology. So if the economics of organic production become unfavourable then the pragmatic producers are more likely to quit this sector. In particular, those attracted to organic farming in the 1990s primarily by price premiums and favourable conversion grants have been the farmers leaving certification as their desired profit margins have failed to materialise. In this respect organic farming is the same as other farming systems: it has to be economically viable in order to survive, and farmers are leaving the organic sector in the same way that they are leaving from other sectors. In this sense organic farming is just another survival strategy, and its economic sustainability has to be in place for it to support an increasing number of producers.

A footnote to this study is provided by those farmers surveyed by Harris *et al*, who commented on the growing involvement of major supermarkets in sales of organic produce, with the perception that this was further reducing farmers' profit margins on their organic output. Some commentators have contended that this supermarket involvement is inimical to the underlying ethos of organic production (Clunies-Ross, 1990). Moreover, Robinson (2004: 237) argues that, whilst supermarkets have been obliged to accept the principles of supplier standards for organic produce based on internationally accepted criteria, some are now introducing their own criteria, which can be applied to their domestic suppliers. In so doing they may champion cheaper alternatives to properly certified organic production, such as produce grown using Integrated Crop Management (ICM), which promotes farmers' reliance on natural predators and crop rotations but does not eliminate the use of artificial fertilisers and pesticides. In 2002 nearly 90 per cent of Sainsbury's UK-sourced fresh produce supplies were being grown to ICM standards, largely through farmers in the LEAF (Linking Environment And Farming) network, which promotes 'green' farming. ICM and Integrated Farming Systems (IFS) address only a certain number of environmental sustainability criteria but exclude others. Hence, Tilzey (2000: 287) concludes that they simply just "attempt to reduce in some measure the ecological 'footprint' of what remains a basically unchanged configuration of intensive, agro-chemically based production."

Genetically-modified (GM) Foods

In the UK especially, the introduction of genetically-modified (GM) foods has been contentious, with strong public suspicion voiced in various ways and government exerting strong pressure to restrict the application of the new technology beyond heavily controlled field trials. Hence, despite the availability of some GM-content foods in British supermarkets from 1995, there were no GM crops being grown in the country at the time of the Glasgow International Geographical Congress in 2004. Bruce Pearce, in Chapter 6, addresses the debate over GM foods in the UK, considering in turn key aspects with respect to environmental, economic and social sustainability.

Of the area currently sown worldwide under GM crops, two-thirds have the principal trait of herbicide tolerance, and the remainder features pest resistance. In both cases farmers are tied to using only that chemical provided by the seed producer. This prevents farmers from moving to cheaper supplies of seeds should they become available whilst providing a lucrative and captive market for the chemical company that supplies the seed. In developing these particular traits and a potentially undesirable relationship for the farmer, the positive claims have been that the overall use of agro-chemicals will be reduced, thereby cutting farmers' costs and lessening exposure of people to herbicides and pesticides. Drawing upon research in the United States by Benbrook (2003), Pearce argues that in recent years the amount of chemicals used on GM crops has actually increased, partly because of growing resistance of weeds to particular herbicides. In terms of environmental impacts, field trials in the UK on oilseed rape, sugar beet and maize have revealed declines in biodiversity for the former two but an increase for the latter, though with concerns about reductions in weeds and weed seeds, and hence potential for longer-term effects on other organisms. In particular it appears that it is the nature of the herbicide regime that is problematic rather than the crops being GM *per se*. Pearce concludes that a move from a conventional to a GM system is likely to have a negative effect on the environment and is therefore highly unlikely to be sustainable. Yet the research does reveal seemingly conflicting findings, as it also does when the economic dimension of sustainability is considered.

The economic advantages of GM crops are postulated as comprising savings on the cost of purchased inputs and increased returns derived from rises in crop yields.

Yet there are contradictory estimates that dispute both the savings and increased yields. So Pearce asserts that, like the environmental impacts of GM crops, the finances need to be assessed on a case-by-case basis. GM crop production does not necessarily reduce financial risk for farmers and, in a climate of consumer resistance, may actually increase risk. In assessing the growth of this resistance in the UK, Pearce discusses the so-called GM dialogue in which the government sought the public's views about GM food and crops. The findings, published in 2003 (DTI, 2003), reported on 675 meetings attended by over 8,000 people, and over 1200 letters and e-mails to the overseeing steering committee. The seven principal conclusions make uneasy reading for supporters of GM. The dominant view of GM was negative, and concerns had grown as people became more engaged with the issues. Over half the participants wanted a ban on the growing of GM crops in the UK and even those in favour of commercial development of GM crops wanted strong regulation. The prevailing negativity reflected a marked mistrust both of government and of the multinational corporations associated with GM technology. However, there was support for further research and greater dissemination of information on GM, especially from 'independent' sources. There were mixed views as to whether Developing Countries represented a 'special case' for which GM technology represented a 'solution' to food production problems.

One of the problems regarding the ongoing GM debate is that it is generally quite narrowly conceived. The GM crops grown around the world have focused largely on more effective control of weeds, with two-thirds engineered for herbicide tolerance. Production of these GM crops is dominated by just a few major chemical corporations that have created crops that are dependent on the chemicals developed by each company, for example Monsanto's creation of plants to be resistant to the high-selling Roundup herbicide. The reliance of farmers upon the chemical companies creates an unhealthy and uneven dependency, with farmers totally reliant on the chemical companies for the chemicals and seeds needed each year. Moreover, it seems to increase reliance on chemical treatments, substituting chemical inputs for labour, and moving farming systems further away from more sustainable alternatives.

The GM dialogue in the UK revealed various other key concerns: the high costs likely to be born by farmers, the possibility that only toxin-resistant plants will survive to reproduce thereby threatening biodiversity, the possible evolution of resistant strains of pests, eradication of a wider range of flora by over-reliance on broad-spectrum weed killers, inter-breeding of species to produce aggressive herbicide-resistant super-weeds, and the 'cascade effect' on the food chain eliminating certain insects at a particular point in the chain by killing their food source. Pearce's chapter highlights that, at least in the UK, the wide extent of public concern over GM foods seems likely to halt the adoption of GM crops on a larger scale at least in the short- to medium-term. Instead the focus of public discourse on food issues has shifted towards greater desire for 'healthy' foods and the need for more consumption of quality foods that can in some respect be linked to health and well-being.

Combining Agriculture and Conservation

In an analysis of current conceptions of post-productivism, Geoff Wilson (2001: 92–5) suggested that ideas about this controversial term could be usefully developed in two ways. One of these was to examine the 'territorialisation' of actor spaces: that is to consider the ways in which both productivism and post-productivism have produced significant spatial variation in the countryside of the Developed World. This variation itself has been conceptualised in various ways, often in the form of other dichotomies, e.g. the 'two-track' countryside (Ward, 1993) and the super-productivism – rural idyll 'spectrum' (Halfacree, 1999). In their simplest terms these dichotomies and spectra differentiate between areas dominated by intensive farming methods, such as East Anglia, the Netherlands, the Paris Basin and Emilia Romagna in a West European context, and agriculturally marginal upland where post-productivism is far more prevalent. However, generally, combinations or variations of the two polar opposites are recognised, as in Marsden's (1999) three rural 'dynamics' (the agri-industrial, the post-productivist and the rural development). All three dynamics may occur at the same time and in close spatial proximity and hence they are not spatially defined. Moreover, it is not implied that particular associated institutional forms, networks, ideologies and norms attached to the different dynamics are superseded in an evolutionary process.

Wilson's second postulate regarding an alternative to the productivist-post-productivist dichotomy builds on this notion of the contemporaneous presence of multiple 'dynamics' or regimes by recognising the existence of multifunctional agricultural regimes (MAR) (2001: 95). He contends that this term is more appropriate than post-productivism in conceptualising changes in contemporary agriculture and rural society, and suggests that, in effect, a MAR "will be characterised by territorialisation of both productivist and post-productivist action and thought, incorporating a multiplicity of responses to the challenges of post-productivism, but with all responses co-existing" (Robinson, 2004: 73). This notion of multifunctionality has been further developed by Wilson (2007) in an extensive analysis of the term's genesis and meaning. He notes the use of the term 'multifunctional agriculture' in connection with the CAP in the early 1990s and in the US Conservation Security Program in 2002. In addition to this policy-related use of the term, there has also been a somewhat limited academic discourse on MAR, focusing on structuralist interpretations of agricultural and rural change (McCarthy, 2005).

The term 'multifunctional agriculture' was first used 'officially' in 1993 by the European Council for Agricultural Law as part of attempts to harmonise agricultural legislation across Europe and to provide a legal basis for sustainable agriculture. However, a specific commitment to multifunctional agriculture appears in the 1996 Cork Declaration by the European Commission. "This Declaration recognised the declining economic role of conventional agriculture in marginal rural areas and the need to find other rationales for public subvention... It also emphasised that agriculture should be seen as a major interface between people and the environment, and that farmers have a responsibility as 'stewards of the countryside'" (Wilson, 2007: 184). Key consequences of this view were that the Second Pillar of the CAP was established and multifunctionality was agreed and introduced in subsequent

EU policy as well as appearing in national agricultural policies at the end of the 1990s (Potter and Burney, 2002). Reforms to the CAP, in the form of Agenda 2000, upheld what was referred to as 'the European model of multifunctional agriculture', essentially recognising the environmental or stewardship role of farmers alongside their role in producing food and fibre whilst also embedding these within wider notions of rural development. Further recent policy-related statements about multifunctional agriculture have appeared from the World Trade Organisation (WTO), emphasising that agriculture produces 'non-commodities' (such as environmental goods), and hence there are concerns beyond those of trade that need to be addressed in policy mechanisms (McCarthy, 2005).

One of the consequences of the engagement of policy-makers with multi-functional agriculture is that the term has been used in an increasingly wide array of different forms during the last decade. This has started to cause confusion about its meaning, not least because there has been little accompanying academic theory to link together the widely varying perspectives on the concept. Wilson (2007) has addressed this, helpfully providing a summary of definitions and seeking to extend the theoretical underpinning. He notes that in its broadest sense multifunctional agriculture is farming that serves multiple functions, thereby reducing the emphasis on production of food and fibre. Therefore this embraces production of environmental goods on farmland, farm diversification and pluriactivity and various related developments, including the strengthening of social capital in the countryside as part of wider improvements in the relationship between agriculture and rural development. However, much of the discourse on multifunctionalism has ignored both the relative weighting of these various components and the extent to which they are interconnected (Wilson, 2007: 186). Interpretations generally seem to focus on economic aspects (e.g. Durand and Van Huylenbroek, 2003), thereby ignoring broader and less tangible factors such as cultural, mental and attitudinal changes as well as complex changes in society-agriculture interactions (Buller, 2004).

In investigating various views of multifunctionalism – economistic, policy-related, holistic, cultural, spatial – Wilson largely agrees with Holmes' (2006: 42) assessment that "the direction, complexity and pace of rural change in affluent, Western societies can be conceptualized as a multifunctional transition, in which a variable mix of consumption and protection values has emerged, contesting the former dominance of production values and leading to greater complexity and heterogeneity in rural occupance at all scales." This emphasises multifunctionality as a territorial concept and acknowledges that agriculture is inherently multifunctional. Within the EU this notion was advanced by Agriculture Commissioner Franz Fischler, who linked agriculture to rural development, tacitly acknowledging that until the late 1990s policy-makers had regarded the two as separate entities.

It is possible to recognise various degrees of multifunctionality: from a very weak development focused on agribusiness and productivism to a strong version embracing conservation measures, agri-environmental policies and farm diversification, especially into production of environmental goods. In policy terms there has been growing support for strong multifunctionality through promotion of sustainable agriculture and farm diversification. In addition, various factors have contributed to the growth of this 'strong' version as illustrated in growing emphasis on short food supply chains, quality foods, extensification, sustainability and 'embeddedness' of agricultural production within an integrated farming/rural community. The ultimate development of this strong version of multifunctionality may be the growth of non-agricultural land use in which land is used in an exclusively non-productivist fashion. Traditionally, such land has been found in areas where farming is characterised by low intensity, high biodiversity and associated with closely integrated rural communities in which agriculture is not necessarily the main source of income. Examples highlighted in the geographical literature include many upland and mountainous areas (e.g. Dax and Hovorka, 2004; Diamond, 2006; Lang and Heasman, 2004; Wilson and Memon, 2005). However, in Chapter 7 Short demonstrates that in the United Kingdom there are other areas that might qualify for the term 'multifunctional' and which exist at the non-productivist interface between agricultural and conservation land use. This raises an important new perspective on the nature of multifunctional agriculture whilst introducing a sustainability dimension into the discussion.

Short focuses on high-value nature conservation sites within lowland England, describing these as pockets of agriculturally marginal land that have largely been abandoned by the spatially more dominant conventional productivist agriculture. Hence these conservation sites are essentially post-productivist, but they also possess the capability of fulfilling multiple functions. Protection of wildlife and natural features takes the highest priority in management terms, but the sites may also possess high levels of amenity value and so they can offer a strong consumption function for people living in the vicinity and for recreationists from a wider area. This amenity potential may have been increased by the additional legal right of public access to such areas created under the Countryside and Rights of Way Act passed in 2000.

Short argues that lowland common land represents the archetypal multiple function land, combining aspects of protection, conservation and consumption (recreational use) with a small amount of productive capacity (usually animal grazing), and with different organisations valuing it for various reasons. With complex legal protections over such land, the commons have remained largely free from the pressures of productivist agriculture and so they are now dominated by recreation, heritage and nature conservation land uses. However, with their multifunctional capacity, the commons require a different approach to nature conservation management than, say, National Nature Reserves or Sites of Special Scientific Interest (SSSIs). This is because of the particular combination of environmental, social, historical and economic interests represented in the commons. Short identifies 'sustainable' strategies that involve community participation (such as the Futures Programme in Scotland), careful management of grazing (termed conservation grazing), and maintenance of historical sites (as in the Countryside Agency's Countryside Character approach).

Based on research for a number of interested agencies, Short identifies various rules for management of multifunctional common land. Essentially these refer to the importance of identifying the key interested parties (stakeholders) and involving them in collaborative decision–making. In particular, the involvement of local communities is increasingly being seen as crucial to maintaining a balance between competing land uses, so that the conservation aspects of management have to embrace local people, as well as landscape, history and access issues. This has the vital ingredient of 'reconnecting' people to their local surroundings within the context of an integrated system of nature conservation and agricultural production. It also offers the prospect of enhancing sustainability across a number of dimensions, and may gain new respect for the role of the lowland commons as part of a model of consensus-based multifunctional land management.

Sustainable Rural Communities

The notion of sustainability when applied to rural communities has tended to be used primarily in terms of the health of a local economy or the preservation of a local culture. Within these two broad areas there may be allied concerns for many separate facets of rural life, including maintenance of social networks, religion, heritage, agriculture and ecology. The sustainability of a rural community may be threatened if the ecosystem that supports its local economy is compromised in some way, but other local factors, such as employment, services and community leadership, need to be in place for a community to sustain itself. So it is quite easy to identify developments that destabilised rural communities in recent decades by rendering them less sustainable. A common scenario has been for industrial-style farming practices not only to yield direct environmental disbenefits but also to produce negative economic and social outcomes for local communities by reducing employment opportunities, fostering out-migration and concomitant reductions in rural services. In general, resource-based economic activities in the countryside have diminished in importance as providers of employment, shedding labour whilst also helping to contribute to the growth in other sectors of the economy that have encouraged progressively greater dependence of the rural population upon urban centres for jobs and services. This has produced a cumulative effect of reducing traditional self-reliance and eliminating closed, more sustainable, systems in rural areas as the globalised economy impacts ever more strongly, even in peripheral regions remote from major urban-industrial centres.

For many rural areas in the Developed World, processes of counter-urbanisation have led to gains in the adventitious population choosing to live in the countryside, but this has not necessarily revitalised rural communities. The majority of in-migrants maintain an urban focus for work, shopping and socialising to create 'dormitory' communities that lack cohesion, services and often possess an imbalanced age structure, as in the case of areas attracting retirement migrants and second-home owners, and those losing young adults to urban centres providing better educational and employment opportunities. This suggests that all dimensions of sustainable rural communities have been eroded by the changing basis of wider economic and social developments. Attempts to prevent the seemingly inexorable demise of rural communities through targeted economic development policies have largely failed, in part because of the now well-recognised limitations to the theories and models on which they have been based (Epps, 2002). Nevertheless, throughout the Developed World there remains a vast array of policies aimed at regenerating rural economies, and with some producing evidence of rural revival. The more significant and most frequently applied mechanisms include changes to regulatory frameworks for business, direct government investment, decentralisation schemes, infrastructure projects, environmental initiatives, education and training, grants, concessions and other forms of direct support. Surprisingly few of these have managed to successfully link economic, social and environmental aspects of rural development, despite certain identifiable successes in individual schemes, e.g. in the form of establishing new businesses, improving infrastructure, maintaining services, and rewarding entrepreneurship and creation of new rural networks.

Of course the impacts of globalisation and the myriad interactions with urban centres do not have to be entirely detrimental to rural communities. For example, the proximity of the urban market may enhance opportunities for rural businesses to meet recreational needs and to supply goods for tourists. New niche markets can be developed, supplying customers travelling from urban areas, e.g. organic foods. Moreover, a declining farm sector may be revitalised by an influx of new part-time or hobby farmers taking over failing farms and shifting the focus of production, as in the case of 'boutique' wineries in New Zealand (Robinson, 1997). In part, differences between positive and negative changes experienced by rural communities reflect not only geographical location but also the extent to which the members of these communities are able to respond positively to the changing circumstances (Bryant, 2002). So local 'actors' are at the heart of translating the various forces impacting on local communities into positive outcomes for the community. If individual actors are able to engender collective actions in particular communities then the ability to produce such positive outcomes can be enhanced. Indeed, it has been argued that it is this dimension of 'collective' action that lies at the very heart of basic principles of sustainable rural community development (Bryant *et al.*, 2001).

Given the pervasive influence of globalisation and urbanisation, it is unrealistic to equate a 'sustainable rural community' with a self-contained and self-sufficient community. However, there can be far greater levels of local reliance across various sectors than has been the case hitherto throughout most of the Developed World. This can apply to housing, economic production, conservation, retailing and other commercial activity. At the heart of this is the creation of economic opportunities so that a sufficiently large base is provided to retain young people within the community and to produce a local tax assessment base to pay for services for local residents. Whilst this represents the operation of the market economy, other functions of sustainability can involve non-market mechanisms relating to conservation of natural and cultural heritage, and the integration of different sets of values within the community into appropriate ways of managing change. This generally involves a series of actors, including individual residents, community groups, real estate agents, property developers, investors, entrepreneurs and government agencies. These different interests have to be successfully integrated into the planning process to secure lasting results for communities. Moreover, this integration must also include adequate consideration of key sectors, such as transport, power generation, employment and waste management, if development is to comprise a sustainable quality. In reality, few plans, other than for very small-scale projects, have been sufficiently wide-ranging and encompassing of either multiple sectors or local actors to generate the necessary conditions for sustainability.

For most rural communities only limited steps have been taken towards producing a sustainable future. Perhaps the most common starting point has been to stabilise the economic base by developing niche land-based production activities combined with local processing for sale to both distant markets and locally. Local markets may be enhanced by tapping into the tourist sector. Tourism itself offers new opportunities, but requires local entrepreneurship and careful management in order to generate local jobs and to prevent leakage of financial returns outside the local milieu. By whatever means, though, jobs need to be created in the rural community to prevent the out-migration of young people and thereby loss of their vitality and skills. In more remote areas this is difficult to achieve if there are no adequate post-secondary educational opportunities available. In areas readily accessible from towns, the difficulty may be in preventing the rural settlement from becoming little more of a dormitory with very little social cohesion.

In addressing the issue of how to generate more sustainable rural communities, Bryant *et al* (2002) stress the need for three priorities to be recognised by different levels of government:

- The importance of local communities in identifying and defining the community's needs.
- The need for sustainable development practices to be 'appropriate', that is taking into account local traditions, cultural values and the general circumstances of specific areas and populations.
- The need to ensure and encourage the integration of broader societal goods and objectives into planning and management for sustainable development.

A key implication here is that national and supranational institutions must emphasise "the creation of enabling frameworks and programmes for planning and managing specific resources, as well as ... communities or groups of communities" (Bryant *et al.* 2002: 273). In this respect the EU's LEADER scheme has the potential to foster sustainable development by promoting inclusive processes of management and change and by building capacity amongst local actors for leadership, entrepreneurship and civic responsibility.

Culture Economy

One approach to the attainment of sustainable rural development has focused on the concept of culture economy. This emphasises opportunities to utilise cultural resources for economic purposes. Hence the distinctive characteristics of local areas and their cultural practices can be developed for commercial purposes, but without endeavouring to apply scale economies to enhance production. This commodifying or valorising of previously non-transacted commodities by local people can be a means of increasing social capital and thereby of empowerment. It is presented in some arguments as the antithesis of development based on external investment, exploitative economic relationships and leakage of profits from the local economy

(Ray, 1998). It is an approach that injects a strong cultural dimension to development alongside economic, social and environmental components, thereby stressing that attempts to attain sustainable development have to embrace the multi-faceted quality of this process.

The culture economy concept accepts that 'culture' comprises a set of resources that can be developed for the economic and social advantage of local inhabitants. It is influenced by the changing nature of post-industrial, consumer capitalism; the trajectory of rural development policy; and the growth of regionalism. It is "primarily concerned with the 'production' side; that is, the territory, its cultural system and the network of actors that construct a set of resources to be employed in the pursuit of the interests of the territory" (Ray, 1998: 4).

Ray (1998) refers to four different modes of culture economy discernible in rural areas:

- Cultures as 'markers' of territorial identity, in which landscape characteristics, language, local tradition and norms confer distinctiveness upon a locality. This can produce a particular image that may be used both formally and informally in promotional and marketing exercises.
- The cohesive use of regional labelling. The construction and projection of territorial identity to people outside the region. This is the 'selling' of places for tourism and inward investment.
- The 'selling' of the region to the communities, businesses, groups and official bodies within the local area. This is intended to stimulate endogenous development, especially in peripheral areas through reinvigorating local culture as a foundation for local and/or regional well-being.
- The various alternative paths of development that a particular region can pursue, in particular stressing local self-reliance in the use of physical resources. The value of local culture is recognised and reinforced.

Cawley and Gillmor (Chapter 8) apply the concept of culture economy to the development of integrated tourism, using the example of western Ireland as a case study. In this instance tourism is viewed as using local resources and local inhabitants in promoting development that draws on local physical, economic, social and cultural environments. Linkages between these environments emphasise local ownership and 'embeddedness' within local structures, whilst a series of local networks can be identified that reinforce endogeneity whilst also possessing links to tourist interests at different geographical scales, from the local to the international. It is argued that embeddedness can confer local distinctiveness that generates a competitive advantage or rent for local products, provided promotion and marketing are effective enough to tap into non-locals so that externally-driven income is injected by tourists into the local economy. An economic multiplier effect can be generated if the tourism is complementary to existing local economic and social structures. However, complementarity may be compromised if the unique character of local culture is jeopardised by tourist development. This parallels one of the central dilemmas of so-called 'sustainable' tourism, in that crucial qualities of a particular environment can easily be compromised by a surfeit of tourist developments.

For 'integrated' tourism to contribute satisfactorily to a particular culture economy, the tourist development has to occur at a scale appropriate to the local resource base. Urry (2002) argues that, typically, this involves locally-owned, small-scale enterprises, such as bed-and-breakfast accommodation and activities based on small groups or even individuals. Direct interaction between tourists and local people is another characteristic feature as is the growth of networking between local enterprises and organisations through partnership formation.

Cawley and Gillmor's analysis focuses on the extent to which tourist developments in part of western Ireland fit this model, in which integrated tourism should be part of a culture economy that empowers local people, delivering benefits to the community and providing a platform for further development. Within the context of an expanding tourist industry in the region, they identify various aspects of complementarity, embeddedness and networking. A related development is the growth of entrepreneurial opportunities for women, enhancing local empowerment. However, their study also reveals a need for a number of refinements to the current 'model' of tourist development if a more multi-faceted sustainability is to be created. In particular, there are differences between the priorities of entrepreneurs and controllers of resources that echo common tensions in many aspects of rural development. Not surprisingly the entrepreneurs have an economic focus whilst the resource controllers show more concern for conservation and the environment. This difference can be seen in some lack of networking between the two groups.

A crucial aspect of any form of economic development in rural areas is the extent of the input from government. For the west of Ireland there have been interventions by the EU, notably in the form of the LEADER (*Liaison Entre Actions de Development de l'Economie Rurale*) programme, which has promoted tourism, and through the reformed CAP, whose second pillar has been supportive of rural development measures. At a regional level the Regional Tourism Authority has channelled investment to tourist businesses and to controllers of resources. There are also state agencies with response-bilities for various aspects of tourism development, e.g. training, and local Chambers of Commerce who can promote certain activities, e.g. networking. Hence, one key conclusion regarding the role of integrated tourism and culture economy in promoting greater rural sustainability is that the role of government at various levels as an 'enabler' is crucial to the outcome of any project.

Farming and Community

One of the major problems affecting the sustainability of rural communities is the parlous financial state of farming in many parts of the Developed World, but especially in areas where agriculture has been the mainstay of the economy and a major source of employment. Lois Mansfield (Chapter 9) focuses on one such area, the uplands of Cumbria in northern England. Although her study area is within Lake District National Park where tourism has become an integral feature of the rural landscape, it is upland agriculture that has produced and maintained many of the attractive features of the landscape. Hence its survival has more than just economic significance. Her analysis describes several of the key landscape components directly attributable to the upland farming system whilst outlining the decreasing economic viability of that system. As returns from traditional sheep and cattle grazing have fallen, farmers have increasingly shed labour from their farms and/or have diversified into tourism-related enterprises or gained additional revenue by participating in agri-environment schemes. However, the numbers of farms in the area continue to decline and along with them the agricultural workforce.

Mansfield's research focuses on one programme, the Cumbria Hill Sheep Initiative (CHSI), intended to reinvigorate the economic potential of the hill farming industry. Funding has come from both local and national sources, but primarily from the EU LEADER+ programme. The CHSI includes a traineeship scheme, to ensure that the farm families have suitably trained offspring who will inherit their farms, and also a social capital research project that is investigating the non-economic value of hill farming, the Fell Farming Traineeship Scheme (FFTS). The FFTS focuses on the training of young people aged between 16 and 30 who have restricted opportunities to enter farming. Each trainee in the scheme is placed with a cluster or 'ring' of farmers representing a varied enterprise mix, though all having hill sheep flocks.

The chapter evaluates the FFTS in terms of trainee development and the labour issue. The scheme delivered strong practical training, though the trainees' theoretical understanding was not especially good. Moreover, those without additional education in a relevant land-based area were largely unable to demonstrate knowledge and practical understanding of environmental issues. This is important because half of the farmers in the scheme regarded understanding of how to work with the unique environment of each farm as crucial to farming success. However, such knowledge and understanding is only acquired with experience, which young trainees cannot really be expected to have.

The second part of this chapter focuses on how the maintenance of an upland farming system can contribute to the production of public goods, arguably a significant contributor to rural sustainability. A secondary aim was to examine the public's perception of the relationship between agricultural activity and public goods. This involved investigation of both the demand and supply sides for public goods covering a range of non-market outputs, including environmental (landscape, habitats), health and safety (animal health, food safety), cultural (rural character, local food) and social (tourism, social cohesion). Samples were taken amongst people living locally and in Manchester. Results from similar surveys were confirmed, in that the Manchester sample regarded geographical qualities of the uplands, such as scenic views and wildlife, as more important than did the residents of the uplands, who stressed 'traditional farm management' and 'community culture'. Responses of those in the sample were subject to a contingent valuation exercise that revealed each household surveyed was willing to pay an average of £47 per annum for upland public goods.

In examining the supply side, the farming community expressed concern at the demise of cooperative behaviour, which had been vital to the management of the flocks of sheep in the uplands. This was affecting the smooth operation of flock management and partly reflected the impact of the declining labour force. The farmers regarded traditional farming skills, the presence of small family farms and a strong local culture as the chief benefits of upland farming. They valued farming's production of peace and tranquillity, and scenic views as the least important products in a list of eight possible benefits. In contrast, the interviewees from Manchester regarded peace and tranquillity as second only to wildlife as the principal benefit. The Cumbrian sample also placed wildlife first, with 'strong local culture' second and 'community culture' third. Clearly there are differences of opinion between the farmers and the other two groups surveyed and this may have important implications for programmes designed to encourage farmers to produce public goods. In particular, they viewed wildlife as an externality to the farming system and landscapes as a functional component of farm management rather than as a 'scenic view'. This essentially productivist view of farmers is at variance with the post-productivist public attitudes.

Despite the prevalence of productivist values amongst the Cumbrian farmers, all the farmers in the FFTS and Social Capital in Hill Farming (SCHF) projects were participating in major agri-environment schemes at the time of the survey, either the Environmentally Sensitive Areas (ESAs) or Countryside Stewardship Scheme (CSS) agreements. It appears therefore that it is the financial gains associated with these programmes that have been the paramount motivation for participation. Any subsequent benefits to the landscape, wildlife and habitat and public would seem to be incidental. In effect, participation in the agri-environment schemes is just one of several income diversification strategies being pursued. One of these strategies highlighted in Mansfield's chapter is the way farmers are building on the commercially attractive properties of hill sheep breeds, such as the Rough Fell and the Herdwick, with diverse product lines to which value is added by use of the breed name or local place names.

In terms of the production of a more sustainable countryside, the challenge in upland regions such as Cumbria is to successfully integrate different components of productivism and post-productivism so that viable communities can be maintained. But it would appear that a fundamental shift has occurred in the way in which the uplands function: from a regime in which environmental goods are a by-product of farming to a regime in which farming is a by-product of environmental goods. Mansfield refers to this as a change from hill farming being a primary producer to it being a tertiary provider. This is being reinforced by the CAP support system moving towards payment of environmental subsidies (i.e. for public goods) rather than payment of subsidies for production of meat and wool (DEFRA, 2005). It remains to be seen, though, just how readily the upland farmers adopt this new 'post-productivist' mindset.

Citizen Participation and Sustainability

In his imaginative geographical perspective on sustainability, Mark Whitehead (2007) perceptively claims that "all forms of sustainability are ultimately local sustainability" (p. 187). The fundamental supporting context underlying this statement is that it is people, both as individuals and as part of wider society, whose actions determine the extent to which any process meets sustainability criteria. Individuals' actions may have an impact at a variety of spatial scales, but in the first instance they occur in a particular locale, often identified as synonymous with a 'local community' (Agnew, 1987). Given the pervasive nature of globalisation the local can never hope to exist in isolation and it is hard to conceive of a confined sustainable local community. Nevertheless, sustainability is often conceptualised as originating at a local community level and its generation has been promoted by various measures of support for 'bottom up' schemes that draw upon local community resources.

This is implicit in the United Nations Agenda 21 programme, which emphasises the need for a move to a more indigenous-based and locally sensitive set of strategies to produce more sustainable communities. It is more explicit in the Local Agenda 21 (LA21) programme launched at the Earth Summit in Rio de Janeiro in 1992. This emphasised the need to involve people actively in the creation of a more sustainable society via consultation and consensus building through which local authorities are then able to benefit fully from the knowledge and ideas contained within the citizenry and in local, civic, community, business and industrial organisations. Therefore the conception of sustainable development involving at its base actions taken by individuals and local communities intersects with the notion of sustainable citizenship in which the individual, through a web of rights and responsibilities, determines just how sustainable any given society actually is. In practice, though, individual actions are mediated by informal and formal political structures, and by interactions between civil society and the state. So individual and local citizenship interacts with the state at various levels, and the actions of individuals and local groups can be supported in multiple ways by the state. It is the extent of control by the state that differentiates between different modes of citizenship and action.

Citizenship can be primarily passive, in which individuals are largely just passive recipients of state policy. This may take the form of individuals using services provided by the state, but not necessarily responding to entreaties that might promote greater sustainability, e.g. recycling waste or reducing industrial carbon 'footprints'. Brown (1997) refers to this as the 'clientalist' mode of citizenship. However, there are various contrasting modes to this passivity, with active citizenship involving the participation of individuals in community activities in order to build on state service provision or compensate for its inadequacies. Such activity may still operate within a broad set of parameters set out by the state, but it may also be both active and radical by challenging the values and practices that are promoted within dominant modes of state-centred citizenship, as in the case of so-called eco-activists who promote alternative systems of environmental rights and responsibilities (Whitehead, 2007: 196). The active, radical approach to citizenship embraces concerns that extend well beyond the local community to include an active interest in the welfare of environments and people who live in other parts of the world, beyond the national boundaries of the citizen's own country. It also embraces a more wide-ranging set of relationships between people and the non-human world than is usually indicated in models of citizenship (Bullen and Whitehead, 2005).

With respect to rural communities there is much scope for far greater consideration of how the various forms of active citizenship can be fundamental to promoting sustainability. Research questions need to be explored relating to the nature of rural social networks, participation of individuals in cooperative ventures, and the roles of rural leadership and entrepreneurship. Some of these issues are addressed by Guy Robinson in Chapter 10, where he discusses two 'environmental' programmes implemented in Canada, one (in Ontario) aimed at promoting environmentally-friendly farming practices and the other (in Atlantic Canada), which encourages local communities to identify key environmental problems and to address them using a combination of local resources and input from provincial and federal agencies.

These two Canadian schemes aim to deliver environmental benefits, both of which include an element of community participation and decision-making 'from below' as opposed to imposition of goals and outcomes 'from above' by government. An inherent element of the schemes is the philosophy that positive environmental outcomes are best delivered if strong citizen participation and even control is integral to the scheme's composition. It is possible to extend this belief into the broader argument that sustainability must necessarily embrace widespread citizen participation and 'bottom-up' control as 'top-down' imposition of solutions is inherently unsustainable. In this sense community participation becomes the 'fourth arm' of sustainability, equally as vital in moves towards attainment of greater sustainability as the economic, ecological and social dimensions of the process. Indeed, it is possible that a lack of participation may substantially restrict the ability of the other three dimensions to produce increased sustainability.

The two schemes evaluated in this chapter are the Atlantic Coastal Action Program (ACAP) and Ontario's Environmental Farm Plan (EFP). Both have operated from the early 1990s and began with a vision and schema from an arm of government. In the case of ACAP this vision was more overt and directed from above, whilst with the EFP it was largely farmers' organisations, in conjunction with input from a provincial ministry, which formulated the scheme's outline and direction. In both cases, though, the 'bottom-up' aspect has been strongly developed over time, giving ample opportunity for the implications of this to be considered.

ACAP began life as a scheme instigated by Environment Canada who set an agenda for environmental actions to be pursued in selected watersheds in Atlantic Canada. The actions were to be delivered through partnerships forged between stakeholder groups who had a vested interest in achieving environmental improvements. At one level it can be argued that everyone has such an interest, but the reality of the stakeholder groups has been that it is key sectoral representatives who have formed the membership of each ACAP round-table: government at various levels, academia, business (and especially those associated with particular environmental problems), the principal economic interests, environmental groups and citizens' organisations. In each of the original 13 ACAP groups a coordinator funded by government was appointed and a facilitator from Environment Canada helped to establish the stakeholder group and relay the initial agenda from which the specific environmental actions were to be delivered. Robinson's analysis, incorporating field-work in three of the more rural ACAP areas, shows that the individual groups formulated comprehensive environmental management plans (CEMPs) that substantially modified and refined Environment Canada's initial objectives. In part this reflected the variable quality of the environmental problems faced across the 13 ACAP groups. Some were experiencing major problems because of sewage discharge into harbours; others had a bigger issue of eutrophication through run-off from farmland;

and others could identify particular polluting industries, such as paper mills, that needed to be tackled. However, it was the variable nature of the decision-making by the individual stakeholder groups that produced such variety and the broad scope of the plans formulated and actions taken.

One of the principal outcomes of the process has been the generation of identifiable actions that would have been most unlikely to have occurred had Environment Canada simply chosen to tackle the problems 'in house'. Through the stakeholder groups, greater citizen awareness of local environmental issues has occurred, with more direct participation in certain environmentally-based activities, e.g. water quality monitoring and clean-up campaigns. Furthermore, control of the agenda has been moved away from the government agency to the stakeholder group and hence closer to the citizenry. This has certainly created a greater sense of 'ownership' of the actions taken and of the associated decision-making, even if the stakeholder groups are not necessarily all accountable directly to a particular constituency. The efforts taken to involve 'polluting' industries have helped to make certain firms confront their environmentally undesirable activities and to play a major role in subsequent amelioration and restoration. Perhaps the most positive overriding outcome of the process, however, has been the removal of decision-making from the sole purview of a government agency so that the 'top-down' dimension has been substantially diluted. This has enhanced sustainability in the sense that, even if future direct financial support for the ACAP scheme by government is removed, there is a strong likelihood that the stakeholder groups will continue to take environmentally desirable actions. Indeed, several of the groups have already demonstrated a substantial degree of financial independence from government by drawing upon private-sector funds to support their activities.

The second environmental scheme examined by Robinson, Ontario's EFP, also has a strong 'bottom up' component, being derived largely from the activities of farmer-based organisations seeking to reduce the worst environmental excesses of 'industrial-style' farming to pre-empt possible future government regulation. This fear of regulation appears to have been an important factor in prompting farmer participation in the voluntary scheme. Other key factors have been farmers' genuine concerns to improve environmental standards on farms, financial incentives (although these have been quite restricted) and peer pressure to be good stewards of the land. The latter can be seen to have been powerful in some areas, with a definite 'neighbourhood effect' operating. However, there has been marked spatial variation in participation, with a definite sectoral bias towards higher levels of participation on livestock farms and quite limited involvement by farmers in the fruit and vegetable sector.

Robinson's analysis examines some of the differences between the EFP and agri-environment schemes operating within the EU. In the terminology employed by Smithers and Furman (2003), the EFP emerges as a 'needs focused' approach, where the actions taken by individual farmers as part of the scheme are determined by their own prioritisation of the environmental actions needed on the farm. Although this choice has input from a Program team comprising peers and representatives from farmers' organisations, it appears to be a less proscriptive arrangement than that operating in major EU agri-environmental schemes. A limitation of the Canadian model is that the selection of environmental actions to pursue on each participating farm generally relates to only particular activities and is therefore not necessarily a whole-farm approach. Moreover, the actions are not intended to promote extensification or any form of lower input-output production; they are intended to ameliorate undesirable disbenefits arising from modern 'industrial' farming. This is in contrast to many of the ideals embodied within EU agri-environment schemes where funding is provided to promote extensification and deliberately to prevent further 'industrialisation' of the farming system so that the aesthetic character of attractive landscapes is preserved and undesirable ecological consequences of farming activities are minimised. The different emphasis in Ontario's EFP means that there is no real intention to move away from the dominant industrial farming model.

In the Ontarian case study the 'bottom up' model was shown to be delivering environmental benefits whilst failing to disturb the dominant characteristics of the prevailing farming system with its dependency on oil-based inputs, large-scale mechanisation and high energy consumption. So the scheme represents only a very minor move in the direction of greater agricultural sustainability.

In broader terms, although the commonly accepted components of sustainability are environmental, economic and social, the Brundtland Report identifies an additional one: institutional arrangements. These can be either a barrier to or a facilitator of sustainable development by promoting or discouraging particular types of behaviour amongst individuals and groups (Spangenberg and Schmidt-Bleek, 1997). Hence rather than dismissing the roles of either Environment Canada or the Ontario Ministry of Agriculture on the grounds that they represent 'authority' and are therefore antithetical to sustainable development that must have a community-led dimension, it is their facilitating role that should be emphasised. In other words, they can represent a favourable institutional dimension that is instrumental in the production of environmental benefits and moves towards greater sustainability. This echoes findings in several of the chapters in both halves of this book.

For rural systems to be truly sustainable there needs to be a partnership between people and government that transcends 'top down' arrangements that have been dominant in Developed Countries. This requires both a change in attitude by those in power as well as the creation of effective mechanisms for involving the widest possible cross-section of individuals within communities. Only then is it likely that there will be policies developed and actions taken that address the range of economic, social and environmental considerations that are embedded within the concept of sustainability. Whilst the concept itself may be open to criticism for its imprecise nature and myriad interpretations, the combination of its three prime dimensions coupled with citizen participation seems to offer the best hope for forging a 'balanced' and viable future for agriculture and rural communities. Unfortunately, as demonstrated by the studies contained herein, we are currently a long way from realising this.

References

Adams, W.M. (1995) Sustainable development? In: Johnston, R.J., Taylor, P. and watts, M. (eds) Geographies of global change: remapping the world in the late twentieth century. Oxford: Blackwell, pp. 354–373.

Agnew, J. (1987) *Place and politics: the geographical mediation of state and society* (Boston: Allen and Unwin).

Allen, P., van Dusen, D., Lundy, L. and Gliessman, S. (1991) Integrating social, environmental and economic issues in sustainable agriculture, *American Journal of Alternative Agriculture*, 6: 34–39.

Arden-Clark, C. and Hodges, R.D. (1988) The environmental effects of conventional and organic/biological farming systems. II Soil ecology, soil fertility and nutrient cycles, *Biological Agriculture and Horticulture*, 5: 223–287.

Bell, D. and Valentine, G. (1997) *Consuming geographies: we are what we eat.* London: Routledge.

Benbrook, C.M. (1990) Society's stake in sustainable agriculture. In: Edwards, C.A., Lal, R., Madden, P., Miller, R.H. and House, G. (eds) *Sustainable agricultural systems*. Arkeny, Iowa: Soil and Water Conservation Society, pp. 37–38.

Benbrook, C.M., 2003. Impacts of genetically engineered crops on pesticide use in the United States: the first eight years, *BioTech InfoNet, Technical Paper*, Number 6, November. http://www.biotech-info.net/Technical_Paper_6.pdf (first accessed Sept 2006).

Boulay, A. (2006) An analysis of farm diversification in France and the United Kingdom based on case studies of Sud Manche and west Dorset, unpublished PhD thesis, School of Geography, Kingston University, London.

Bowler, I.R. (2002a) Developing sustainable agriculture. *Geography*, 87: 205–212.

Bowler, I.R. (2002b) Sustainable farming systems. In: Bowler, I.R., Bryant, C.R. and Cocklin, C. (eds) *The sustainability of rural systems: geographical interpretations*. Dordrect: Kluwer, pp. 169–188.

Bowler. I.R., Bryant, C.R. and Cocklin, C. (eds) (2002) *The sustainability of rural systems: geographical interpretations*. Dordrecht: Kluwer Academic Publishers.

Bryant, C.R. (2002) Urban and rural interactions and rural community renewal. In: Bowler. I.R., Bryant, C.R. and Cocklin, C. (eds) (2002) *The sustainability of rural systems: geographical interpretations*. Dordrecht: Kluwer Academic Publishers, pp. 247–270.

Bryant, C.R., Allie, L., Des Roches, S., Buffat, K. and Granjon, D. (2001) Linking community to the external environment: the role and effectiveness of local actors and their networks in shaping sustainable community development. In: Pierce, J.T., Prager, S.D. and Smith, R.A. (eds) *Reshaping of rural ecologies, economies and communities*. Burnaby, British Columbia: Department of Geography, Simon Fraser University, pp. 57–67.

Byant, C.R., Bowler, I.R. and Cocklin, C. (2002) Conclusion. In: Bowler. I.R., Bryant, C.R. and Cocklin, C. (eds) (2002) *The sustainability of rural systems: geographical interpretations*. Dordrecht: Kluwer Academic Publishers, pp. 271–274.

Bullen, A. and Whitehead, M. (2005) Negotiating the networks of space, time and substance: a geographical perspective on the sustainable citizen, *Citizenship Studies*, 9: 499–516.

Buller, H. (2004) The 'espace productif', the 'théâtre de la nature' and the 'territoires de développement local': the opposing rationales of contemporary French rural development policy. *International Planning Studies*, 9: 101–119.

Buttel, F.H., Larson, O.F., Gillespie Jr. G.W. (1990) *The sociology of agriculture.* Westport, CT: Greenwood Press.

Callicott, J. and Mumford, K. (1997) Ecological sustainability as a conservation concept. *Conservation Biology*, 11: 32–40.

Carley, M. and Christie, I. (2000) *Managing sustainable development.* London: Earthsan, second edition.

Charlton, C.A. (1987) Problems and prospects for sustainable agricultural systems in the humid tropics. *Applied Geography*, 7: 153–174.

Clunies-Ross, T. (1990) Organic food: swimming against the tide. In: Marsden, T.K. and Little, J.K. (eds), *Political, social and economic perspectives on the international food system.* Aldershot: Avebury, pp. 200–214.

Cobb, R. Dolman, R. and O'Riordan, T. (1999) Interpretations of sustainable agriculture in the UK. *Progress in Human Geography*, 23: 209–235.

Dawson, J. and Burt, S. (1998) European retailing: dynamics, restructuring and development issues. In: Pinder, D. (ed.) *The new Europe: economy, society and environment.* Chichester: John Wiley and Sons, pp. 157–176.

Dax, T. and Hovorka, G. (2004) Integrated rural development in mountain areas. In: Brouwer, F. (ed.) *Sustaining agriculture and the rural environment: governance, policy and multifunctionality.* Cheltenham: Edward Elgar, pp. 124–143.

Department of the Environment, Food and Rural Affairs (DEFRA) (2005) *Environmental Stewardship.* London: Rural Development Service.

DEFRA (2007) United Kingdom milk prices http://statistics.defra.gov.uk/esg/datasets/milkpri.xls (first accessed Sept. 2006).

Department of Trade and Industry (DTI) (2003) *GM Nation? The findings of the public debate.* Crown Copyright. http://www.gmnation.org.uk/docs/gmnation_finalreport.pdf (first accessed Sept 2006).

Diamond, J. (2006) *Collapse: how societies failed to succeed.* London: Penguin.

Doering, O. (1992) Federal policies as incentives or disincentives to ecologically sustainable agricultural systems. *Journal of Sustainable Agriculture*, 2: 21–36.

Dunlap, R.E., Beus, C.E., Howell, R.E. and Waud, J. (1992) What is sustainable agriculture? An empirical examination of faculty and farmer definitions. *Journal of Sustainable Agriculture*, 3: 5–40.

Durand, G. and Van Huylenbroek, G. (2003) Multifunctionality and rural development: a general framework. In: Van Huylenbroek, G. and Durand, G. (eds) *Multifunctional agriculture: a new paradigm for European agriculture and rural development.* Aldershot: Ashgate, pp. 1–18.

Epps, R. (2002) Sustainable rural communities and rural development. In: Bowler. I.R., Bryant, C.R. and Cocklin, C. (eds) (2002) *The sustainability of rural systems: geographical interpretations.* Dordrecht: Kluwer Academic Publishers, pp. 225–246.

Evans, N.J., Morris, C. and Winter, M. (2002) Conceptualizing agriculture: a critique of post-productivism as the new orthodoxy. *Progress in Human Geography*, 26: 313–332.

Fairweather, J.R. (1999) Understanding how farmers choose between organic and conventional production: results from New Zealand and policy implications. *Agriculture and Human Values*, 16: 51–63.

Foster, C. and Lampkin, N. (2000) *Organic and in-conversion land area, holdings, livestock and crop production in Europe*. Report to the European Commission, FAIR 3-CT96-1794, University College Aberystwyth.

Francis, C.A. and Younghusband, G. (1990) sustainable agriculture: an overview. In: Francis, C.A., Flora, C.B. and King, L.D. (eds) Sustainable agriculture in temperate zones. New York: John Wiley & Sons, pp. 1–23.

Giddens, A (1998) *The third way*. Cambridge: Polity Press.

Goodman, D. (2003) Editorial: the quality 'turn' and alternative food practices: reflections and agenda, *Journal of Rural Studies*, 19 (1): 1–7.

Halfacree, K.H. (1999) A new space or spatial effacement? Alternative futures for the post-productivist countryside. In: Walford, N.S., Everitt, J. and Napton, D. (eds) *Reshaping the countryside: perceptions and processes of rural change*. Wallingford: CABI, pp. 67–76.

Hamilton, S. (2001) Impacts of agricultural land use on the floristic composition and cover of a box woodland in Northern Victoria. Pacific Conservation Biology, 7: 169–184.

Holmes, J. (2006) Impulses towards a multifunctional transition in rural Australia: gaps in the research agenda. *Journal of Rural Studies*, 22: 142–160.

Lang, T. and Heasman, M. (2004) *Food wars: the global battle for mouths, minds and markets*. London: Earthscan.

Lockie, S. (1998) Landcare and the state: action at a distance. In: Burch, D., Lawrence, G., Rickson, R.E. and Goss, J. (eds) *Australia's food and farming in a globalised economy: recent developments and future prospects*. Clayton, Vic: Monash Publications in Geography and Environmental Science, no. 50.

Lowrance, J. (1990) Research approaches for ecological sustainability. *Journal of Soil and Water Conservation*, 45: 51–57.

McCarthy, J. (2005) Rural geography: multifunctional rural geographies – reactionary or radical? *Progress in Human Geography*, 29(6): 773–782.

Marsden, T.K. (1999) rural futures: the consumption countryside and its regulation. *Sociologia Ruralis*, 39: 501–520.

Mercer, C. (2006) Low milk prices unsustainable, say producers, *Dairy Reporter. com*, 11.5.06.

Ministry of Agriculture, Fisheries and Food (MAFF) (2000) *England Rural Development Plan*. London: MAFF.

National Farmers Union (NFU) (2007) British milk – what price 2007? http://www.nfuonline.com/x15301.xml (first accessed May 2007)

Nygard, B. and Storstad, O. (1998) De-globalization of food markets? Consumer perceptions of safe food: the case of Norway. *Sociologia Ruralis*, 38: 35–53.

O'Riordan, T. and Cobb, R. (1996) That elusive definition of sustainable agriculture. *Town and Country Planning*, February, pp. 50–51.

Potter, C. and Burney, J. (2002) Agricultural multifunctionality in the WTO: legitimate non-trade concern or disguised protectionism? *Journal of Rural Studies*, 18: 35–47.

Ray, C. (1998) Culture, intellectual property and territorial development, *Sociologia Ruralis*, 38 (1), 3–19.

Robinson, G.M. (1997) Farming without subsidies: lessons for Europe from New Zealand? *British Review of New Zealand Studies*, 10: 89–104.

Robinson, G.M. (2002a) Nature, society and sustainability. In: Bowler. I.R., Bryant, C.R. and Cocklin, C. (eds) (2002) *The sustainability of rural systems: geographical interpretations*. Dordrecht: Kluwer Academic Publishers, pp. 35–58.

Robinson, G.M. (2002b) Guest editorial: Sustainable development – from Rio to Johannesburg. *Geography*, 87(3): 185–188.

Robinson, G.M. (2004) *Geographies of agriculture: globalisation, restructuring and sustainability*. Harlow: Pearson.

Robinson, G.M. and Ghaffar, A. (1997) Agri-environmental policy and sustainability, *North-West Geographer*, third series, 1: 10–23.

Robinson, G.M., Loughran, R.J. and Tranter, P. (2000) *Australia and New Zealand: economy, society and environment*. London: Arnold.

Sachs, W. (1999) *Planet dialectics: explorations in environment and development*. London: Zed Books.

Smithers, J. and Furman, M. (2003) Environmental farm planning in Ontario: exploring participation and the endurance of change, *Land Use Policy*, 20: 343–356.

Sneddon, C.S. (2000) 'Sustainability' in ecological economics, ecology and livelihoods: a review, *Progress in Human Geography*, 24: 521–549.

Spangenberg, J.H. and Schmidt-Bleek, F. (1997) How do we probe the physical boundaries for a sustainable society? In: Ryden, L. (ed.) *Foundations of sustainable development: ethics, law, culture and the physical limits*. Uppsala: Uppsala University Press.

Tilzey, M. (2000) Natural areas, the whole countryside approach and sustainable agriculture. *Land Use Policy*, 17: 279–294.

Troughton, M.J. (1993) Conflict or sustainability: contrasts and commonalities between global rural systems. *Geography Research Forum*, 13: 1–11.

Troughton, M.J. (1997) Scale change, discontinuity and polarization in Canadian farm-based rural systems. In Ilbery, B.W., Chiotti, Q. and Rickard, T. (eds) *Agricultural restructuring and sustainability: a geographical perspective*. Wallingford: CAB International, pp. 279–292.

Troughton, M.J. (2002) Enterprises and commodity chains. In Bowler, I.R., Bryant, C.R. and Cocklin, C. (eds), *The sustainability of rural systems: geographical interpretations*. Dordrecht: Kluwer Academic Publishers.

Trzyna, T. (ed.) (1995) *A sustainable world: defining and measuring sustainable development*. IUCN/Earthscan: London.

United Nations (2002) Johannesburg Summit Brochure, http://www.johanes burgsummit.org/html/brochure (first accessed 5 June 2002).

Urry, J. (2002) *The tourist gaze*. London: Sage.

Ward, S. (1993) The agricultural treadmill and the rural environment in the post-productivist era. *Sociologia Ruralis*, 33: 348–64.

Whitehead, M. (2007) *Spaces of sustainability: geographical perspectives on the sustainable society*. Routledge: London and New York

Whiteside, M. (1998) *Living farms: encouraging sustainable smallholders in South Africa*. London: Earthscan.

Wilson, G.A. (2001) From productivism to post-productivism ... and back again? Exploring the (un)changed natural and mental landscapes of European agriculture. *Transactions of the Institute of British Geographers*, new series, 26: 103–120.

Wilson, G.A. (2007) *Multifunctional agriculture: a transition theory perspective.* Wallingford, Oxon and Cambridge, Mass: CABI.

Wilson, G.A. and Memon, A. (2005) Indigenous forest management in 21st century New Zealand: towards a 'post-productivist' indigenous forest-farmland interface, *Environment and Planning A*, 37: 1493–1517.

World Commission on Environment and Development (WCED) (1987) *Our common future.* Oxford: Oxford University Press.

Yarwood, R. and Evans, N.J. (2000) Taking stock of farm animals and rurality. In: Philo, C. and Wilbert, C. (eds) *Animal spaces, beastly places.* London, Routledge, pp. 98–114.

Young, W. and Welford, R. (2002) *Ethical shopping: where to shop, what to buy and what to do to make a difference.* London: Vision Paperbacks.

Chapter 2

Productivism versus Post-Productivism? Modes of Agri-Environmental Governance in Post-Fordist Agricultural Transitions

Mark Tilzey and Clive Potter

Introduction

Much debate concerning the sustainability of agriculture in industrialised countries relates to the extent to which it has become 'post-productivist' (see, for example, Wilson, 2001; Evans *et al.*, 2002; Walford, 2003). The way in which this debate has been configured appears, however, to display a number of shortcomings: it both fails to conceptualise how sustainable agriculture might be defined and implanted in broader social relational terms and, by the same token, fails to adduce causality underlying the 'forms of production' that these analyses seek to describe. At root, these shortcomings appear to derive from the current tendency to counter-pose local agency against exogenous 'structural force' and a related tendency to conflate the concepts 'post-productivism' and 'post-Fordism' (see Potter and Tilzey, 2005 for extended discussion). The first fails to consider how local agency is articulated with wider structures of material and discursive power. The second, by conflating post-productivism, a form of production, with post-Fordism, a regime of accumulation/ mode of social regulation, fails relatedly to assess the degree to which the regulatory regimes that encompass forms of production are undergoing transition.

This chapter attempts to overcome these shortcomings by, firstly, conceptualising initiatives to secure sustainability ('post-productivism') as variously concordant with, subordinate, or oppositional with respect to the hegemonic regime of accumulation. Where concordant with, or subordinate to the prevailing regime of accumulation, sustainability initiatives may become part of a new mode of social regulation. These 'systemic' variants (comprising 'weaker' forms of sustainability) are either functional with respect to the prevailing accumulation dynamic (for example, supplying quality products to the 'reflexive' middle-class consumer) or are palliative/mitigatory, being directed towards selective amelioration of its environmental and social diseconomies (Tilzey, 2002). 'Anti-systemic' variants (comprising 'stronger' forms of sustainability) tend to problematise the dominant accumulation model as itself the primary author of environmental contradictions, advocating a reformation of social relations such that human needs are addressed equitably within the parameters defined by the capacities of biophysical nature. Social movements advocating such an approach (for example, *Via Campesina*) are therefore anti-systemic with respect to the dominant growth paradigm.

Secondly, the chapter attempts to comprehend the configuration of social modes of regulation by reference to a class based and political ecological ontology. The nature and balance of class interests within, and at the level of, the state appear to be a vital determinant of the character of sustainability and degree to which it becomes part of a social mode of regulation. The deployment of the concept 'class' here seeks to capture the structural character of the distribution of interests in society and their reproduction through agency. Class as 'structured agency' is thus operational at all spatial scales, vitiating the dichotomy between 'behaviourally' grounded explanations at local level and those grounded in 'structure' at higher and wider scalar levels. A political or social ecological perspective, meanwhile, sees nature and society as internally related, rather than as discrete entities (see Burkett, 1999; Castree, 1995; 2005; Tilzey, 2002). Whilst non-human nature is, in varying degrees, socially (re)constructed in material terms through human manipulation of ecosystems and the genetic material of ecosystem agents, such 'hybrid' agents nevertheless retain materiality and agency, thereby supplying affordances and imposing constraints on human actions on nature. Combining these two approaches, we can discern how accumulation dynamics implicate environmental changes and unsustainable outcomes. Nature, now re-conceptualised as 'social' nature, is necessarily integral to the study of change in agri-food systems both in terms of the way it is an inherent element of the production process and in terms of the way it is impacted upon sustainably or unsustainably by that production process. Nature is thus part of the materiality of uneven development in agriculture. And more reflexively, of course, the environmental contradictions contingent on accumulation processes implicate the differential integration of nature as the object of social modes of regulation by states.

The dynamic of capitalism(s) describes the underlying rationale of politico-economic change. Capitalisms, instantiated in the form of the state, variously take cognisance of environmental contradictions, selectively mitigating impacts but seeking to minimise adverse implications for growth and profitability, or rendering them in such a form (for example, ecological modernisation) as to enhance accumulation. In short, states will seek to enhance material sustainability to the extent that this is consistent with maintaining the relational sustainability of capitalism (see Drummond and Marsden, 1999). It is contended then that the nature and depth of sustainability initiatives in state policy largely conform to the economic governance 'norm-complex' defined by the dominant regime of accumulation (see Bernstein, 2002). Thus, almost irrespective of the nature and severity of environmental contradictions or the effectiveness of proposed responses, state environmental policy will tend to reflect the dominant governance norms underpinning capital accumulation. While reactions to environmental (and social) contradictions in civil society may assume both anti-systemic and more systemic forms, the institutionalisation of environmental response in state policy is predicated on the systemic promotion and maintenance of the prevailing regime of accumulation. The prevailing regime of accumulation is determined primarily by the balance of class interests in the state and by the latter's status within the 'world system' – whether core/dominant or peripheral/subordinate.

These factors, then, are primary determinants of how environmental contradictions are configured in social consciousness and how the 'norm-complex' structures state governance of the environment. Considerations relating to the nature and severity of environmental contradictions themselves appear to be of less significance in determining the mode of state environmental governance.

These observations have important implications for our understanding of the character and causal basis of post-productivism. First, post-productivism needs to be understood within the context of broad social relational changes related to the profound shifts in the pattern of agricultural governance and modes of regulation that have emerged since the 1980s – the transition from Fordism to post-Fordism. The neo-liberal governance norms of post-Fordism have presented an increasing challenge to the 'exceptionalist' status of agriculture under Fordism (see Tilzey, 2006) and entail a 'de-centring' of agriculture in those spaces where global competitiveness is difficult to sustain through productivist scale economies. The neo-liberal challenge has been accompanied by environmental critiques of productivism, related in turn to the 'second contradiction' of capital (see O'Connor, 1998). The resulting selective demotion of productivism has presented opportunities for the implantation of post-productivism, particularly in economically marginal and disadvantaged areas. Productivism remains characteristic in economically advantaged spaces, now assuming, however, an increasingly 'market productivist' form (see Potter and Tilzey, 2005; Tilzey, 2000). In contrast to much of the literature addressing the productivism – post-productivism issue, which has tended to construct an either/or dualism, we can perceive, therefore, that, in their totality therefore post-Fordist spaces tend to *juxtapose* a dominant market productivism and a subordinate post-productivism. Second, post-productivist initiatives, whether anti-systemic or not in social origin, tend to undergo 'systemicisation' as they pass through the 'norm-complex' filter in order to secure embodiment in state policy. As state policy, post-productivism has become in many senses part of the systemic processes of 'denationalisation' and 'destatisation' (see Jessop, 2002) that form key elements of post-Fordism. The first entails, *inter alia*, the devolution of regulatory responsibilities for post-productivist initiatives to the regional and local levels but without proportional transfers of power or capacity (McCarthy and Prudham, 2004). The second involves increasing levels of public-private partnerships in the design and delivery of policy and a concomitant reconfiguration of policy according to the neo-liberal norms of entrepreneurialism and 'value for money'.

The current politico-economic conjuncture is thus defined by a dominant shift in the regime of accumulation from Fordism to post-Fordism. Post-productivism, as but one – and a relatively subordinate – facet of this transition, expresses the selective decline of productivist agriculture under conditions of globalising competition and responses to the environmental (and social) contradictions of productivism. State level, systemic responses to post-productivism, however, exhibit variability in the degree to which they are willing to intervene in the market to secure sustainability objectives within a context given by the ascendancy of processes of neo-liberalisation (Larner, 2003; Peck and Tickell, 2002; Tilzey, 2006). This variability appears to co-vary according to the depth of neo-liberal policy implantation in the economy. High levels of neo-liberal retrenchment in the economy tend to generate low levels of market intervention in environmental governance norms. More qualified acceptance of neo-liberalism in economic policy tends to coincide with more interventionist proclivities in the governance of post-productivism. This chapter proposes to explore the nature and causal bases of this variability in the governance of post-productivism with respect to three developed capitalist polities – the European Union (EU), Australia and the USA. In the light of Castree's related and detailed discussion of 'neo-liberalising nature' (Castree 2007a; 2007b), we should perhaps note that the processes of neo-liberalisation analysed in this chapter appear to be primarily of a *sui generis* kind, particularly in the case of Australia, while being reinforced at the same time, in the case of the EU and USA especially, by trans-national sources of neo-liberal governance, flowing from the World Trade Organisation (WTO) for example. Further, variability in neo-liberalisation appears to flow not so much from differences in the form of neo-liberalism in each instance, its internal discursive consistency apparently being maintained in each, but rather from the way in which neo-liberalism is subject to articulation with, and compromise by, other class and related interests within each state-society complex. Further variability in governance norms is introduced by the particular agential character of the ecological base in each of the three polities examined. Finally, some commentary on the relative effectiveness of these governance modes in relation to sustainability criteria is also offered.

Post-Productivism in the European Union (EU)

Post-productivism in the EU appears to conform to an 'embedded neo-liberal' mode of governance (Potter and Tilzey, 2005; Tilzey, 2006; Tilzey and Potter, 2007). This reflects the coincidence of strong pressures for market liberal restructuring with mitigatory impulses flowing from the continued political influence of anti-free market middle/small farm constituencies, a geo-politically powerful polity with resilient social democratic instincts, and traditionally robust relations of joint production between agriculture and biodiversity and landscape. These class, geo-political and ecological characteristics constitute key elements defining agricultural and environmental policy in the European transition from Fordism to post-Fordism.

Under Fordism, political productivism assumed a concordance between the accumulation (economic) and legitimation (socio-political) functions of policy (Bonnano, 1991), a homogeneity built on the premise that agricultural production would automatically reproduce the socio-cultural qualities of rural space (Gray, 2000). This policy homogeneity was built in turn upon a nationally centred, Keynesian domestic order providing for a compromise between capital and labour in the wider economy. In the 1980s, however, there emerged in Europe an increasing regional differentiation between those areas with an increasing predominance of larger, capitalised farms (primarily of low relief and fertile soils) and those in which smaller farms remained dominant (primarily of difficult topography and poorer soils). Simultaneously, the environmental impacts of political productivism began to enter social consciousness as traditional agri-ecosystems, particularly in the lowlands, were lost or degraded through production intensification and specialisation. Where, previously, there had been an actual or assumed symmetry between agriculture and socio-cultural/natural values, there now emerged a spatial asymmetry, with economic functions (productivism) being emphasised in some spaces, and social, cultural and environmental functions (post-productivism) in others.

In response, the policy homogeneity of Fordism began to exhibit fragmentation as economic functions failed increasingly to coincide with the social (and now more explicitly environmental) aims of policy. Thus, from 1985 the EU began to pursue a modified ('embedded') neo-liberal strategy through the creation of a single market and the move towards economic and monetary union, on the one hand, and through an emphasis on 'cohesion', on the other (Coleman, 1998; van Apeldoorn, 2002). The Fordist image of the farmer as *producer* of food for the *nation* began to concede – at least in policy discourse and as a policy goal – to the image of the farmer as *entrepreneur* active in a *global* economy. Reflecting the continuing economic and political influence of neo-mercantilist and social protectionist constituencies, however, this emergent neo-liberal regime of accumulation was to be significantly modified through the provision of social safety nets and a variety of 'catch up' supports. This emergent policy set functioned both to facilitate accumulation in the new post-Fordist mould and to mitigate its adverse impacts through social safety measures, now legitimated increasingly through adhesion to cultural and environmental functions.

In this way, both the structural and price/market dimensions of agricultural policy entered a long, contested and still current process of reform to accommodate the new heterogeneous vision for farming in which there were to be two basic forms of agriculture. The first, a restructured, commercial agriculture, was now expected increasingly to compete unaided in the global economy, while the second, a social (cohesion) agriculture, was now re-conceptualised as 'agriculture-in-a region'. Whilst the new post-Fordist paradigm emphasises agriculture as a sector like any other – the de-legitimation of economic 'exceptionalism' in accordance with neo-liberal norms – it also stresses for 'social cohesion' agriculture, however, an image of farming as contributing to 'regional and rural development'. This constitutes the re-legitimation of social 'exceptionalism' on the basis of agriculture's *multifunctions* – social, cultural, environmental – in certain regions. At the same time, the 'new environmentalism' legitimates certain forms of agricultural subvention, as the environmental role of farming in sustaining valued habitats and landscapes is seen to be fundamental to the new status of (selected) rural areas. Traditional social support is thus re-legitimated through rural development and environmental functions, serving to placate both rural social and urban environmental interest groups – a fusion of interests rendered feasible by relations of joint production between extensive farming and positive environmental externalities in European rural space.

This reform agenda, articulated and promulgated by the European Commission, reflects in no small measure the greatly increased influence of trans-national, neo-liberal class interests in defining and promoting a more globally and market-oriented agricultural policy (Potter and Tilzey, 2005; Tilzey and Potter, 2007; van Apeldoorn, 2002). These interests are also defining the parameters within which agri-environmental policy is formulated and, increasingly, the very content of that policy as it becomes subject to neo-liberal forms of environmentalism (Castree 2007a; McCarthy and Prudham, 2004). These influences are subject to qualification, however, with policy taking the form of 'embedded' neo-liberalism, juxtaposing market productivism and post-productivism as 'integrated rural development' (CEC, 1996). This strategy appears designed to stimulate the further expansion of productivism, now of a more market-oriented kind, and its increased integration into the agri-food circuits of capital. The progressive elimination of 'market distorting' support in Pillar 1 of the EU's Common Agricultural Policy (CAP) is complemented by the creation of Pillar 2, with the intention to afford some measure of continuing support to farms most marginalised in this process, to provide countryside consumption spaces for the urban populace (whilst conserving a residual biodiversity and landscape resource), and to supply the middle-class 'reflexive' consumer. A bi-polarity in policy is therefore evident between globalising norms of governance for market productivism and regionalised norms of governance for 'multifunctional' agri-rural activities. The result is likely to be an enhanced duality in rural space between a dominant market productivism and a subaltern and marginalised post-productivism, the whole subordinate to a neo-liberal discourse of competitiveness and entrepreneurialism.

The process of CAP reform thus bears the clear imprint of post-Fordist modes of governance. The overarching rationale underlying reform is embodied in the process of 'denationalisation' (the selective rescaling of regulatory functions to supra- or sub-national levels), as the WTO, for example, increasingly imposes neo-liberal norms of legitimacy on the direction and content of policy. The CAP is thus becoming progressively more market oriented, a trend manifest initially in price support reductions and direct payments under the MacSharry and Agenda 2000 reforms, and subsequently in further price reductions and the decoupling of direct support under the 2003 Mid-Term Review (MTR). The cumulative character of these reforms has been designed to facilitate the progressive penetration of market relations and market dependency into European agriculture (Potter and Tilzey, 2005; Potter and Tilzey, 2007; Wood, 2002). Meanwhile, the regionalisation or localisation of certain elements of policy is also evident, particularly in relation to Pillar 2. This reflects the 'territorialisation' of policy objectives and delivery in Pillar 2 particularly, in which locality is seen to be a key desideratum in enhancing value-adding in economies of 'scope' and in defining environmental benefits. A concomitant of territorialisation is the trend towards 'destatisation' – the blurring of traditional boundaries between state and civil society – again particularly marked within Pillar 2, as the state seeks to enrol a variety of partners in the definition and operationalisation of objectives for economic diversification.

Particularly in states such as France and Germany, Pillar 2, as the pre-eminent site of 'post-productivism' and agri-environmental policy within the CAP, continues to embody traditions of agrarianism and social protectionism, premised on market interventionism. However, as, and if, reform proceeds along its current trajectory, policy governance in relation to 'post-productivism' appears set to assume an increasingly neo-liberal complexion. Rural Development Regulation (RDR) budgets are likely to be increasingly regionalised in their administration but tightly disciplined and disbursed on a competitive and selective basis, thereby heavily constrained in their ability to counteract overarching processes of restructuring. Budgets for agri-environmental management are likely to be defined and defended increasingly according to neo-classical public goods criteria, entailing more restrictive forms of subvention in line with WTO 'green box' disciplines and the requirement to minimise 'market distortion' (Tilzey, 2006). In complementary fashion, rural development assistance will facilitate less competitive farmers to diversify away from agriculture or render existing enterprises more competitive, while transitional adjustment assistance will smooth the exit of marginal farm enterprises from the industry. Indeed, the recent agreement on a new RDR for the next programming period (2007–2013) has been designed explicitly to conform to the neo-liberal tenets embodied in the EU's Lisbon Strategy, emphasising the priorities of higher economic growth, job creation and greater competitiveness in world markets (CEC, 2005). Therefore pre-eminent amongst the priorities of the new RDR will be assistance for people to adapt to more market-oriented agriculture and the promotion of new ways of selling and dealing with risk in competitive markets.

Agri-environmental and rural development policy in the EU thus appears to be subordinated increasingly to the neo-liberal 'norm-complex'. This is qualified, however, for reasons of social legitimacy and the evident dependency of much biodiversity and landscape quality upon continued agricultural management. In other words, resistance or opposition from class and interest groups centred around neo-mercantilist, social protectionist and environmental discourses have moderated the turn to market productivism. The resulting 'embedded neo-liberalism' comprises a strategy which juxtaposes market productivism and a limited set of agri-environmental and rural development measures to foster 'post-productivism'. The implication is clearly one of a 'zoned' countryside, with large areas of intensive, environmentally poor, but commercially competitive farming, plus marginal and upland zones where farms exist primarily as tools for environmental management, supported by targeted schemes, or as suppliers of niche markets (Tilzey, 1998). The latter measures, however, are unlikely to compensate for falling incomes of small and medium-sized farms, particularly as, despite the introduction of the Single Farm Payment, producers confront a secular fall in commodity prices as liberalisation proceeds (Tilzey, 2005b). Given the joint relationship between farming and nature, together with the increasingly restrictive, public goods character of RDR subvention, it is difficult to envisage how environmental quality at the landscape scale can be sustained as agriculture's principal income source – the sale of commodities – continues to erode.

It is within this context that a putatively oppositional paradigm of post-productivism (more specifically that of endogenous or agrarian-based rural development) has been promoted, premised on the assertion that the market power of corporate food interests can be countered by exploiting the turn by consumers away from industrial food provisioning in favour of quality food production (Marsden, 2003; Marsden and Sonnino, 2005; Morgan *et al.*, 2006; Potter and Tilzey, 2007). While these authors place emphasis on elements – localism, ecological sustainability – that are key to strong sustainability, their paradigm remains centrally wedded to market dependency (see Wood, 2002; 2005) and therefore subject to the contradictions that attend this condition. Thus, the turn to economies of scope and niche markets, and therefore dependency on middle-class consumption as the principal revenue stream for smaller producers, is likely to afford only temporary respite from the pressures of competition as more producers enter the field of quality production. Downward pressure on prices and capital concentration are predictable outcomes, while the volatility and arguably unsustainable nature of upper income consumption – premised as this increasingly is on global, neo-liberal circuits of finance capital – would suggest considerable caution in relation to the longer-term viability of this 'alternative' paradigm. Indeed, these authors (Morgan *et al.*, 2006, p. 195) have themselves expressed reservations concerning the assumed 'alterity' of their paradigm, intimating that the turn to the 'local', when allied to continuing market dependency, may merely represent the 'inside' of a wider process of rescaling the state, the 'outside' being the growth of supra-national scales of governance intimately associated with neo-liberal globalisation. This reprises our earlier discussion of denationalisation and destatisation as key elements of neo-liberalisation, suggesting that while endogeneity and the 'alternative agricultural food networks' paradigm do contain oppositional (anti-systemic) elements, their discourse of market dependency assures tendential conformity to a systemic (neo-liberal) form of post-productivism.

Stronger discourses of 'post-productivism' and sustainability are clearly being marginalised in this process (see Potter and Tilzey, 2005). The realisation of agrarian, non-productivist (see CPE, 2001) and 'whole countryside' perspectives (see Tilzey, 2000), for example, appears an ever more distant prospect, premised as these are on the co-evolutionary principle of 'jointness' in agricultural production and nature throughout rural space. The realisation of a whole countryside vision of this kind, predicated on integrated principles of environmental and resource sustainability, food security and social sustainability, would seem to require the kinds of policy intervention and support that are considered increasingly illegitimate under neo-liberal norms.

Post-Productivism in Australia

Post-productivist initiatives in Australia are configured very largely in accordance with a 'radical neo-liberal' mode of governance. This reflects the hegemony that market liberal policies have attained over the last twenty years and, in contrast to the EU (and US, see below), the virtual absence of mitigatory impulses involving more than a minimum of market intervention. Rather, responses to the contradictions of productivism conform very much to post-Fordist norms of 'denationalisation' and 'destatisation'. This radical neo-liberal mode of governance reflects, in turn, Australia's semi-peripheral status geo-politically and its traditional export dependency. Relatedly, and in contrast to Europe and the US (see below), it also reflects the relative failure of the small and medium farm constituency to differentiate itself politically and ideologically from the large farm constituency. Further, it reflects a recent and predominantly productivist agricultural presence which has militated against the development of strong co-evolutionary relations between farming, biodiversity, landscape and natural resource use.

Notwithstanding Australia's traditional historical status as a dependent economy within the world system, the relatively brief post-war geo-political conjuncture typified by state-centred economies did permit the Australian state to adopt a number of agricultural policy measures that could be described as 'Fordist' in character. The political productivism of this period – the 1950s and 60s – generated a polarising effect on farm structure. Large farms grew increasingly wealthy, while a significant number of medium and smaller farms experienced increasing cost-price pressures and struggled to meet debt repayments incurred during this expansionary phase (Gray and Lawrence, 2001; Lawrence, 1987). In this way the social contradictions of productivism became increasingly evident in the 'plight of the small farmer' and it became clear that expansionary policies were incapable of resolving the 'low income' problem. Similarly, the environmental contradictions of productivism mounted progressively, involving overgrazing, soil erosion and salinisation, loss and degradation of native biotopes, and pollution and depletion of surface and ground water resources (Lawrence *et al.*, 1992).

These contradictions took place against a backdrop in which Australia, by the late 1960s, had moved away from Britain as a trading partner, with the US becoming Australia's largest supplier of imported goods and Japan its largest market. From the late 1960s Australia experienced increasing penetration by US and Japanese capital and, under this globalising influence, Australia's manufacturing industry was construed by an increasingly foreign ownership to have failed to develop successfully as an export sector (Lawrence, 1989; 1990). These forces coalesced to generate strong pressures for a re-alignment of the Australian economy as a supplier of cheap agricultural and mineral products in the emergent post-Fordist global economy. Within this new context Australia, as a high-cost labour region, was seen to be more suited to the production of raw materials, a role that was to place pressure on the primary industries (agriculture and mining) to attain the necessary overseas income to help sustain the nation's metropolitan style of living. The implication was one of thoroughgoing export orientation and the achievement of high productivity/efficiency (capitalisation), premised centrally on the externalisation of environmental and social costs in those primary industries (Lawrence, 1990; Lawrence *et al.*, 1992).

These events signalled a shift in agricultural policy by the state. Thus, by the late 1960s governments felt increasingly that they could no longer maintain assistance based on farmers' costs of production. Policies designed under the Fordist era to keep 'unviable' and 'low-income' farmers on the land – primarily legitimating functions but enabled by the state-centred character of Fordist accumulation – were now perceived by proponents of the ascendant neo-liberal paradigm as counter-productive, inhibiting the emergence of a more internationally competitive farm sector (Pritchard, 2000; 2005a). With the progressive removal of output-based supports and protectionist policies from the 1970s, the state redirected and increasingly confined intervention and subvention to rural adjustment and restructuring (Higgins, 1999). Under new pressures to attain international 'comparative advantage', the old arguments for government intervention to correct market failure in the rural economy suffered de-legitimation. 'Efficiency' and 'competitiveness' became the shibboleths of the new paradigm. Rural adjustment towards increased market orientation was seen as an essential policy objective and, to this end, the Rural Adjustment Scheme (RAS) was introduced in 1977 to assist farmers in 'adjusting' to free market determined prices, rather than relying on previous supply-based output support (Gray and

Lawrence, 2001: Pritchard, 2000). Despite the introduction of a new RAS in 1985, the provisions of the 1977 Act remained essentially intact until 1988 when the RAS was extensively restructured, an event that finally signalled the full implantation of neo-liberal policy and discourse in Australian rural political economy.

The 1988 Act re-oriented the RAS towards improved farm management and productivity and away from family assistance. The reform coincided with a renewed emphasis on agricultural restructuring with a view to supplying cheap food to Asian-Pacific nations, a trend that further stimulated the development of a corporate-based, highly capitalised, deregulated and vertically-integrated farming model (Lawrence and Gray, 2000; Pritchard, 2000). The co-occurrence of RAS restructuring and renewed state encouragement of agri-business is unlikely to have been coincidental since the former was imperative if the state was to send the correct 'signals' to the market by redistributing resources to those (larger) farms and agri-business firms that could deploy them in the most 'efficient' manner. In conjunction with the evolution of rural adjustment legislation, the 1980s saw the Australian state implement a wider range of neo-liberal policies within agriculture and the wider economy (Gray and Lawrence, 2001; McMichael and Lawrence, 2001; Pritchard, 2000). Such measures included: substantial reduction in tariffs, currency flotation and the suspension of foreign exchange controls, deregulation of finance and banking, the adoption of monetarist policies to control inflation, the abandonment of statutory marketing, the removal of direct subsidies to farming, elimination of price supports, dismantling of a favourable tax regime for farmers, severe reduction in government-sponsored research activity, and the introduction of user-pays principles for extension and information services. While the political productivism of the Fordist era generated socio-economic and environmental contradictions, these have exhibited exponential increase under the market productivism of the neo-liberal era. Economically, neo-liberal policies have engendered an intensification of the cost-price squeeze as input prices and interest rates have continued to rise, while commodity prices have slumped due to global oversupply and continuing export subsidisation by the EU and the US, reinforcing pressures towards farm capitalisation and 'efficiency'.

The environmental consequences of the drive to produce food and fibre as 'efficiently' as possible under conditions of high farm capitalisation and cost-price squeeze include the accelerated 'mining' of soils, overstocking/grazing of pastures, dryland and wetland soil salinisation, over-abstraction of meagre water resources for irrigation, pollution of ground and surface waters, further erosion of native biotopes, and overall increased resource intensity of production (Burch *et al.*, 1992; Drummond *et al.*, 2000; Lawrence *et al.*, 1992). In the livestock sector, for example, attempts to increase livestock production have led not only to the widespread loss and degradation of native grassland but also to considerable adverse impacts on soil structure, nutrient cycling, water quality and quantity, pasture productivity and palatability, and thus on the long-term sustainability of the sector itself (Hamilton, 2001; Prober and Thiele, 1995). These contradictions have stimulated the emergence of a range of 'post-productivist' initiatives over the last ten or fifteen years intended variously to address or to mitigate the adverse environmental impacts of productivism. Again, in the livestock sector, research and farm practice have focused primarily on: increasing the drought tolerance and persistence of pastures (Kemp and Dowling, 2000); reducing soil and nutrient loss (Greenwood and McKenzie, 2001), acidification (Kemp and Dowling, 2000), and the effects and extent of dryland salinity; preventing or minimising invasion by low palatability or toxic plants. As a result, there is now increasing recognition in the sector of the contribution of native perennial plant species to livestock, and particularly to sustainable livestock, production in Australia (see for example, Garden *et al.*, 2000), a view contrasting starkly with attitudes of previous decades. The implication is that the retention of biodiversity, together with other components of the natural resource base, is crucial if the longer-term sustainability of livestock production is to be assured. Further, it is increasingly evident that, reciprocally, biodiversity in native grasslands is dependent upon the presence of low intensity and low-input grazing systems (Fensham, 1998; Freidel and James, 1995). The startling implication is that the sustainability of livestock production in Australia appears to require the development of *joint* production relations between farming, biodiversity, and other components of the natural resource base in a manner comparable to the more extensive pastoral systems in Europe, for example.

By the 1990s this recognition of the severity of both biodiversity loss and the functional crisis of agriculture and of the solutions to them had stimulated a re-focus on the need to address 'off-reserve' biodiversity and natural resource conservation (Figgis, 2005; Williams, 2004). However, the way in which these 'post-productivist' initiatives have been configured conforms very much to neo-liberal norms of governance in agricultural and environmental policy. As environmental discourse has expanded from its 'reserves' focus ostensibly to encompass the 'whole countryside', so has it become re-configured and subordinated to a neo-liberal environmental norm-complex. These norms are founded on a number of premises. First, it is held that the contradictions of export-oriented, market productivist agriculture are outweighed by the benefits – Australia derives net 'welfare' gains from the erosion of its rural ecological and social infrastructure. Second, due to these putative gains, fiscal rectitude and the sanctity of free enterprise, policy should not impact centrally on the commercial 'efficiency' of the farm enterprise, nor should it in any way be seen to be subsidising production. Third, the costs entailed in implementing sustainable resource use and any other aspects of management which contribute to the long-term viability of the farm should be borne by landholders themselves, not by the public purse. Public subvention should be confined, therefore, to paying for environmental services of a 'public goods' character. And fourth, public policy should be directed to changing attitudes and management styles (not economic imperatives), to primarily field-edge and palliative actions, and to supporting community and group initiatives rather than individual enterprises (Tilzey, 2005a).

The introduction of National Landcare in 1989 constituted the primary Federal government response to the environmental crisis besetting Australian agriculture, a scheme designed to assist farmers in mitigating problems of biodiversity loss and resource degradation. Symptomatically, Landcare has been configured as a predominantly voluntary and community-based initiative, with government funds disbursed to these groups, not to the farm enterprise. It is legitimated through forms of 'bounded' democratisation and participatory rhetoric and deployed at localised, economically deregulated rural sites (Martin and Ritchie, 1999). While the spatial frame of reference for these mitigatory actions may be strategic, the latter generally fail to address the core economic drivers of degradation, located as these are at the heart of the farm enterprise. Where funds are disbursed to individual enterprises, this is on a strictly public goods rationale, abjuring interference with 'market forces'. The state elevates the fiction of the 'universal' entrepreneur as the effective arbiter of substantive action to address sustainability, the state itself being electively impotent in this arena – another means by which to legitimate inaction. Landcare exemplifies the post-Fordist processes of 'denationalisation' and 'destatisation' in their most unadorned form. As Martin (1997: 53) notes:

> neo-liberalism represents a type of 'degovernmentalization of the state' where state deliberation focuses more towards those mechanisms which enhance self-regulation (the market and local participation), rather than considering the substantive issues facing society. The assumption ... is that local participation within an increasingly de-regulated market environment will produce adequate forms of "incentivised" personal conduct for a sustainable and productive rural sector. Through the production of the calculative, economically rational farmer, environmental problems can be "internalized", while those of a wider spatial and temporal dimension can be captured in the Landcare net.

This attempted normalisation of market-based relations seeks to dissolve the tension between economic and environmental policy, but this fiction can be sustained only if substantive progress in addressing the agri-environmental crisis is forthcoming. While Landcare has undertaken much peripheral and palliative activity, there is little evidence that the core problems of land degradation are being addressed (DAFF, 1999). Whether the recently introduced Natural Heritage Trust can succeed where Landcare has failed remains to be seen, but this new Federal scheme is likely to be configured along similar lines, with monies being disbursed to groups (primarily Catchment Management Boards) rather than directed to farm enterprises (Tilzey, 2005a).

Post-productivist initiatives have thus emerged in Australia since the 1980s within a conjuncture given by the demise of Fordism and the 'second (environmental) contradiction' of capitalism. These initiatives, however, are either internally constituted, or externally constrained, by a hegemonic, neo-liberal norm-complex. As in the EU and the US, a bipolarity in rural space is evident between market productivism and post-productivism, although in the Australian case it is characterised by an overwhelming asymmetry in favour of the former. This pattern of reconstituted and uneven development, in which some parts of Australia may withdraw from agriculture, others experience reinforced (market) productivism, and still others engage in more extensive, 'sustainable' farming for the reflexive consumer, imply that authors such as Holmes (2002) and Argent (2002) are both correct in their analyses, depending upon which facet of this post-Fordist conjuncture one chooses to emphasise. What does seem clear, however, is that post-productivist, environmental initiatives throughout Australian rural space are conditioned profoundly by the arbitration of the market as the fundamental determinant of this pattern of uneven development. Thus, it is those areas on the economic and ecological margins (and remote from urban markets), where competitive scale economies are difficult to realise, that are most likely to withdraw from agriculture (Holmes, 2002). In the productivist heartlands, post-productivist initiatives are confined to field edge and palliative measures lest competitiveness be compromised. Only in those peri-urban and littoral areas, 'each with its own market motif and *cuisine de terroir*' (Argent, 2002: 111), is agriculture likely to undergo an, albeit tenuous, shift to more sustainable forms of post-productivism as market premia here enable an 'internalisation of environmental externalities' that elsewhere would require public support that the state is unwilling to underwrite.

Agri-environmental governance in Australia thus largely ignores the imperatives of accumulation and farm survival that embody the hegemony of radical neo-liberalism. These imperatives, however, constitute the primary drivers of environmental (and social) unsustainability in rural Australia (Lockie and Bourke, 2001). Agri-environmental governance is effectively subordinate to the radical neo-liberal 'norm-complex'. National Landcare is symptomatic of the form and fate of post-productivist initiatives in Australia. Despite its relative popularity amongst farmers, Landcare has proven largely ineffectual in addressing Australia's rural environmental crisis, primarily because its radical neo-liberal configuration denies the need, or capacity, to confront the structural bases of that crisis (Cocklin, 2005; Drummond *et al.*, 2000; Tilzey, 2005a). Under strong pressure to maximise foreign exchange earnings, pursue fiscal 'rectitude', minimise the risk of capital flight, and compounded by the absence of farm constituencies ideologically predisposed to confront neo-liberalism, the Australian state is unwilling to engage in the necessary measures and expenditure to secure sustainability. And yet a profound shift towards sustainable, lower input – lower output, post-productivist agriculture in Australia would appear to constitute an environmental, economic and social imperative (see for example McIntyre *et al.*, 2004). The measures required to bring about this shift, however, including 'positive coordination' of the market (see Tilzey and Potter, 2006), are currently held to be wholly illegitimate according to the tenets of radical neo-liberalism.

Post-Productivism in the United States (US)

The US may be described as something of a hybrid between the 'embedded neo-liberalism' of the EU and the 'radical neo-liberalism' of Australia. With the EU it shares a strong discursive differentiation among farm constituencies, with small and medium-size farm organisations markedly critical of neo-liberal trends and supportive of continued agricultural exceptionalism. With Australia it shares the overwhelming dominance of productivism – although of a more political rather than a purely market kind – in which relations of jointness between agriculture and nature are poorly developed both materially and discursively. This relationship conforms largely to the 'negative impact' model in which the production of nature is conceived as an essentially non-agricultural phenomenon. In contrast to Europe, therefore, the environment has not generally been invoked in the service of agricultural exceptionalism.

The US is the hegemonic 'core' capitalist state and its administrations must seek legitimacy primarily amongst domestic constituencies, in marked contrast to states such as Australia (see above). Additionally, the US was perhaps the epitome of the nationally 'articulated', Fordist state to which family-farm based agricultural production for the home market was central. Like Europe, but unlike Australia, the US thus has a strong tradition of agricultural subvention on both economic and social 'exceptionalist' grounds. Under Fordism agricultural policy attempted, without overwhelming contradiction, to sustain the family farm, whilst simultaneously supplying mass urban markets. Over time, however, productivism engendered considerable farm restructuring, leading to severe erosion in the number of medium- and smaller-size farms engaged primarily in agricultural activity (Buttel, 1989; 2003; Lobao and Meyer, 2001). Under post-Fordism, commercial viability is now secured increasingly via agro-exports which are produced overwhelmingly by very large and large farms, and decreasingly by the upper-middle farm constituencies. Their productive activities are increasingly integrated into trans-national agro-food commodity circuits (Buttel, 2003; McMichael, 2003). Nevertheless, productive activity is still undertaken predominantly on family farms and, as a legacy of the strong corporatist arrangements of the Fordist era, family farm interest groups retain considerable political power within Congress (see for example Moyer, 2004). In varying degrees US administrations must continue therefore to seek political legitimation through placation of these interest groups. Legitimation under post-Fordism is now sought principally through agro-export drives, and the episodic subvention required to secure US competitiveness is in turn legitimated, disingenuously, by references to the vulnerability of the American family farm (Dixon and Hapke, 2003). The primary aim of the US in the international arena with respect to its agriculture sector is therefore to expand foreign markets for its farm commodities, to facilitate corporate and non-corporate accumulation and, thereby, to enhance political legitimacy with respect to these interest groups. This strategy has the additional merit of reducing the budgetary burden of support on the state. US administrations continue to seek, therefore (either willingly or under duress), to reserve the right to subvene farmers by various means largely as a result of the enduring power of class fractional lobby groups within Congress (Moyer, 2004). Discursive legitimation for such subvention is sought by recourse to social and economic 'exceptionalist' arguments, their content nuanced according to the fractional interests of the two major farm constituencies (Dixon and Hapke, 2003), the primarily large-farm Republican (represented by the major commodity groups and the American Farm Bureau) and middle-farm Democrat (represented principally by the National Farmers' Union), respectively.

To the extent that US agriculture is construed within these two hegemonic discourses (Republican neo-liberalism or market productivism and Democrat neo-mercantilism/social income support or political productivism) to be at all 'multifunctional' or 'post-productivist' (terms not yet deployed self-referentially by US interests), the socio-economic dimension is likely to be invoked most strongly in the cause of agricultural 'exceptionalism'. The environment, for its part, has not hitherto been construed as a principal means to justify public support for agriculture *per se*, both because farming is highly productivist (generating the actual or tendential externalisation of nature from production) and, relatedly, because key environmental elements such as biodiversity and landscape either exist, or are perceived to exist, essentially outside and in opposition to agricultural practice (see Bohmann *et al.*,

1999). Thus, as the mass commodity production sector has continued the process of intensification and specialisation (and the occlusion of nature) (see Ritchie and Ristau, 1986; Kenney *et al.*, 1991; Buttel, 1997), so, in tandem, have environmental programmes focused not on the integration of nature with production or the inherent contradictions of the productivist model, but rather on the retirement of lands from production. In this way, the bulk of expenditure on environmental programmes in agriculture, since their inception under the 1985 Farm Bill, has been directed to farmland retirement under the Conservation Reserve Program (Zinn, 1999) and more recently to mitigating the pollution impacts of productivism.[1] In this way US policy appears to represent a hybrid of European and Australian circumstances, marrying the political economy of European interventionism to the political ecology of Australian 'new worldism'.

Agri-environmental policy under the most recent (2002) Farm Bill exhibits a similar profile, with the bulk of expenditure devoted to conservation reserve and pollution mitigation. It likewise represents subordinacy to, and consistency with, the dominant accumulation dynamic of productivism. Nevertheless, the introduction of the Conservation Security Program (CSP) under the 2002 Farm Bill may constitute a new departure for US policy, bringing it closer to a European political ecology in which jointness is key. The CSP would appear to be significant, firstly because it is a programme that disburses funds for '*on-farm*' or working lands practices designed to enhance environmental (and socio-economic) sustainability *through* production (Hoefner, 2003). It therefore represents an incipient shift away from the prevailing negative impact model in US agri-environmental policy towards recognition that post-productivist farm practices have the capacity to generate positive externalities. Secondly, the CSP is significant since its particular form as an agricultural land use payment reflects the emergence of new 'post-productivist' farmer constituencies, located primarily in the Northeast US, together with the new 'reflexive' consumption of sustainably produced food and 'consumption countryside', primarily by urban consumers (see Cowan, 2002). In this way US policy appears to exhibit an incipient, albeit highly asymmetrical, duality with post-productivist payments for 'multifunctional' agriculture supporting the often pluriactive, marginal farmer in areas primarily outside the zones of mass (productivist) food commodity production. This 'post-productivist', pluriactive farmer constituency appears to be most accurately represented politically and discursively by the Sustainable Agriculture Coalition (Hoefner, 2003).

Despite its status as a mandatory 'entitlement' programme under the 2002 Act with funding through the Commodity Credit Corporation – implying the legal entitlement of all farms meeting entry criteria to enrol in the scheme, with funding 'floating' to the level of participation – the current administration and the United States Department of Agriculture (USDA) has thus far succeeded in imposing a strict budgetary cap on the CSP. As a consequence, to date the CSP has been deployed as a spatially constrained and highly discretionary programme that has fallen far short of its real potential (see Molnar *et al.*, 2005). This constriction and marginalisation of the CSP appears to reflect primarily the concern of the current administration to service the Republican farm constituency and the orthodox agri-environmental policies that are concordant with its accumulation needs, but also the prevalence of neo-classical economists within the USDA and their antipathy towards entitlement programmes (ERS, 2003). This is unfortunate since the CSP's focus on environmental conservation *on* working land has the potential to be a viable alternative to reliance on conventional, productivist support, particularly in the case of medium- and smaller-size farms. These farms, particularly those more reliant on conventional programme support, would welcome a scheme in which payments are calculated not merely on environmental costs/values but on income foregone. The CSP is significant, therefore, in being related to levels of income support deemed requisite to continued farm viability. It constitutes an attempt to tie social income and environmental supports to on-farm, post-productivist management practices. The CSP appears to have the potential to combine the benefits of 'broad and shallow' and 'narrow and deep' agri-environmental schemes in its attempt, at least within the upper two tiers of the programme, to generate multi-functions jointly through an agrarian based post-productivism (Sustainable Agriculture Coalition, 2003).

The CSP is a programme which reflects very strong traditions of exceptionalism and market interventionism in policy and is one which could have arisen only in such circumstances. Its political ecological profile, however, is one that does not accord with the continuing prevalence of productivism in the US and its attendant negative impact model of agri-environmental governance. The CSP's current marginalisation appears symptomatic, therefore, of the dominance in US agricultural policy of two hegemonic discourses – radical neo-liberalism and neo-mercantilism. For both, the environmental dimension of policy is one normatively to be pursued off-farm, placing minimal constraints on the pursuit of productivism. While the wider US economy has fallen under the increasing influence of neo-liberal fractions (represented politically chiefly by the Republican Party) since the 1980s in the transition to post-Fordism, fractions of agricultural productive capital have resisted, or at least qualified, these trends largely by successful manipulation of their tactical importance to the continuing hegemony of the ruling party (Moyer, 2004; Orden, 2002). The implication is that, despite its radical neo-liberal rhetoric in the international arena, domestically the US is obliged to pursue a strategy of qualified neo-liberalism. The cornerstone of US strategy internationally is enhanced market access in order to wean its agro-exporters away from market distorting support, both to safeguard its World Trade Organisation Aggregate Measure of Support ceiling and to conform to increased budgetary stringency at a time of economic downturn (Fynn, 2003; Orden, 2003; Petit, 2002). It also wishes to facilitate the globalising strategies of US agro-food multinationals. The feasibility of so doing, however, depends crucially on the compliance of Congress. The US visited decoupled payments following the 1996 FAIR Act and found them wanting in the face of declining global agricultural commodity prices (Orden *et al.*, 1999). Under political pressure from agricultural productive fractions, counter-cyclical payments were introduced under the 2002 Farm Bill to mitigate the impact of such price downturns, constituting a reversal of the trend towards neo-liberal policies. So, while significant elements of the neo-liberal 1996 Act remain in place, the 2002 Act has seen a re-assertion of market interventionism. US agriculture thus continues to be characterised by strong elements of political productivism and an agri-environmental policy which is very much consistent with, and subordinate to, its accumulation imperative. The CSP, although having the potential to deliver huge benefits for economic, environmental and social sustainability, has suffered marginalisation because its principal advocates comprise a farm constituency with little political power. Within the current US post-Fordist conjuncture we may thus discern an extremely asymmetrical duality comprising two hegemonic productivisms (neo-liberalism and neo-mercantilism), aligned with a concordant set of agri-environmental programmes, and juxtaposed to a subaltern, and still very much emergent, form of post-productivism which, significantly, has the potential to conjoin the production of food with biodiversity, landscape and resource conservation.

Conclusions

The case studies presented in this chapter have served to highlight a number of issues surrounding the productivism – post-productivism debate. First, there has been no general shift from productivism to post-productivism – rather there has been a shift to post-Fordism, combining an increasingly dominant market productivism with subaltern elements of post-productivism. This asymmetrical bipolarity is a feature of all the case studies. Second, within this transition to post-Fordism, the particular configurations of agricultural and agri-environmental policy in the polities examined appear to be explicable by reference to: the character and distribution of class and class fractional power in the state-society complex (van der Pijl, 1998); the location of the state within the hierarchy of the 'world-system'; and to the specificities of political ecology within the case studies. Third, it seems evident that initiatives to engender sustainability and post-productivism are subordinate to, or defined by, the dominant regime of accumulation in all three polities. The implication is that stronger models of sustainability are either marginalised, or transmuted, into various weaker models as, and if, they become embodied in state policy to constitute elements of modes of social regulation, a process designed to secure the 'relational sustainability' of capitalism(s). This marginalisation and transmutation is to be much lamented since it continually re-imposes relations of non-integration between the economic, environmental and social dimensions of sustainability, thwarting post-productivism's considerable potential, where embodied in more anti-systemic discourses, to engender the necessary *synergistic* relations between these elements.

Note

1 The 1985 Farm Bill was significant in that for the first time in US agricultural policy the 'environment' became defined, and was to be addressed, as a discrete issue (as opposed to being merely coincidental with acreage reduction programmes, although the latter did enable environmental issues to be accorded policy prominence). Productivism had generated severe environmental problems in terms of accelerated soil erosion, pollution and eutrophication of surface and ground water through agro-chemical inputs, greatly restricted ecological genetic diversity, human and ecological health problems, and the direct destruction or degradation of ecosystems through conversion to arable via ploughing or drainage.

All these impacts had accelerated during the 'boom' years of the 1970s but, significantly, it was the first and last of these only, ie those that were coincident with the interests of acreage reduction, which received real attention under the 1985 Act. Thus, the Conservation Reserve Programme offered farmers assistance to divert highly erodible cropland to grassland or trees, whilst the Highly Erodible Lands sub-title embodied a compliance provision requiring all farmers with highly erodible land to draw up and implement a conservation plan in order to qualify for continued public subvention. Such measures could be employed to placate environmental concerns at least in some degree, but it is surely no coincidence that these measures served simultaneously to reduce deficiency payments directly by reducing production and indirectly by raising commodity prices and thereby reducing the deficiency payment disbursed on remaining crop production.

Likewise, under the 1996 Farm Bill, conservation 'gains' were very much subordinate to the dominant economic motive underlying the drive to reduce the size of the federal deficit. In conservation terms this Act was broadly regressive as cropland was brought into (essentially productivist) production from land previously set aside under the 1985 Act. Receipt of the new decoupled payments under the Act was subject to conservation compliance, but its provisions were somewhat relaxed. Thus, conservation measures under this Act had essentially the same political ecological profile as those of the 1985 Act, although their implementation appears to have been even more contingent upon opportunities opened up by, and consistent with, the dominant accumulation dynamic.

References

Argent, N. (2002) From Pillar to Post? In search of the post-productivist countryside in Australia, *Australian Geographer*, 33 (1), 97–114.
Bernstein, S. (2002) *The compromise of liberal environmentalism*. New York: Columbia University Press.
Bohmann, M. et al. (1999) *The use and abuse of multifunctionality*. Washington DC: USDA Economic Research Service.
Bonanno, A. (1991) The globalization of the agricultural and food system and theories of the state, *International Journal of Sociology of Agriculture and Food* 1, 15–30.
Burch, D., Rickson, R. and Annels, R. (1992) The growth of agri-business: environmental and social implications of contract farming, In Lawrence, G., Vanclay, F. and Furze, B. (eds), *Agriculture, environment and society*. Melbourne: Macmillan, pp. 259–279.
Burkett, P. (1999) *Marx and nature: a red and green perspective*. Basingstoke: Macmillan.
Buttel, F. (1989), The US farm crisis and the restructuring of American agriculture: domestic and international dimensions, In Goodman, D. and Redclift, M. (eds), *The international farm crisis*. Basingstoke: Macmillan, pp. 46–83.

Buttel, F. (1997) Some observations on agro-food change and the future of agricultural sustainability movements, In Goodman, D. and Watts, M. (eds), *Globalising food: agrarian questions and global restructuring*. London: Routledge, pp. 344–365.

Buttel, F. (2003) Continuities and disjunctures in the transformation of the US agro-food system, In Brown, D. and Swanson, L. (eds), *Challenges for rural America in the twenty-first century*. University Park, Pennsylvania: Pennsylvania State University Press, pp. 177–189.

Castree, N. (1995) The nature of produced nature, *Antipode*, 27, 12–48.

Castree, N. (2005) *Nature*. London: Routledge.

Castree, N. (2007a) Neoliberalizing nature: the logics of de- and re-regulation, *Environment and Planning A*, forthcoming.

Castree, N. (2007b) Neoliberalizing nature: processes, effects and evaluations, *Environment and Planning A*, forthcoming.

Commission of the European Communities (CEC) (1996) *The Cork Declaration: a living countryside*. Cork: Report of the European Conference on Rural Development.

CEC (2005) *The Common Agricultural Policy and the Lisbon Strategy*. Brussels: European Commission.

Cocklin, C. (2005) Natural capital and the sustainability of rural communities, In Cocklin, C. and Dibden, J. (eds), *Sustainability and change in rural Australia*. Sydney: University of New South Wales Press, pp. 171–191.

Coleman, W. (1998) From protected development to market liberalism: paradigm change in agriculture, *Journal of European Public Policy*, 5, 632–651.

Coordination Paysanne Européenne (CPE) (2001) *To change the Common Agricultural Policy*. Brussels: CPE.

Cowan, T. (2002) *The changing structure of agriculture and rural America: emerging opportunities and challenges*. New York: Novinka.

Department of Agriculture, Fisheries and Forestry (DAFF) (1999) *Managing natural resources in rural Australia for a sustainable future: a discussion paper for developing a national policy*. Canberra: Department of Agriculture, Fisheries and Forestry.

Dixon, D. and Hapke, H. (2003) Cultivating discourse: the social construction of agricultural legislation, *Annals of the Association of American Geographers*, 93, 142–164.

Drummond, I. and Marsden, T. (1999) *The condition of sustainability*. London: Routledge.

Drummond, I., Campbell, H., Lawrence, G. and Symes, D. (2000) Contingent or structural crisis in British agriculture? *Sociologia Ruralis*, 40 (1), 111–127.

Economic Research Service, United States Department of Agriculture (USDA) (2003), unpublished interview, Washington DC.

Evans, N.J., Morris, C. and Winter, M. (2002) Conceptualizing agriculture: a critique of post-productivism as the new orthodoxy, *Progress in Human Geography*, 26, 313–332.

Fensham, R. (1998) The grassy vegetation of the Darling Downs, South-eastern Queensland, Australia. Floristic and grazing effects, *Biological Conservation*, 84, 301–310.

Figgis, P. (2003) The changing face of nature conservation: reflections on the Australian experience, In Adams, W. and Mulligan, M. (eds), *Decolonizing nature: strategies for conservation in a post-colonial era*. London: Earthscan, pp. 197–219.

Freidel, M. and James, C. (1995) How does grazing of native pastures affect their biodiversity? In Bradstock, R., Auld, T., Keith, D., Kingsford, R., Lunney, D. and Silvertsen, D.(eds), *Conserving Biodiversity: Threats and Solutions*. Sydney: Surrey Beatty and Sons and NSW National Parks and Wildlife Service, pp. 249–259.

Fynn, J. (2003) unpublished interview. Geneva, World Trade Organization.

Garden, D. et al. (2000) A survey of farms on the Central, Southern and Monaro Tablelands of NSW: management practices, farmer knowledge of native grasses, and extent of native grass areas, *Australian Journal of Experimental Agriculture*, 40, 1081–1088.

Gray, I. and Lawrence, G. (2001) *A future for regional Australia: escaping global misfortune*. Cambridge: Cambridge University Press.

Gray, J. (2000) The Common Agricultural Policy and the re-invention of the rural in the European Community, *Sociologia Ruralis*, 40, 30–52.

Greenwood, K. and McKenzie, J. (2001) Grazing effects on soil physical properties and the consequences for pastures: a review, *Australian Journal of Experimental Agriculture*, 41, 1231–1250.

Hamilton, S. (2001) Impacts of agricultural land use on the floristic composition and cover of a box woodland in Northern Victoria, *Pacific Conservation Biology*, 7, 169–184.

Higgins, V. (1999), Economic restructuring and neo-liberalism in Australian rural adjustment policy. In Burch, D., Goss, J. and Lawrence, G. (eds), *Restructuring global and regional agricultures: transformations in Australasian agri-food economies and spaces*. Aldershot, Ashgate, pp. 131–143.

Hoefner, F. (2003) unpublished interview. Washington DC: Sustainable Agriculture Coalition.

Holmes, J. (2002) Diversity and change in Australia's rangelands: a post-productivist transition with a difference?' *Transactions of the Institute of British Geographers*, new series, 27 (3), 362–384.

Jessop, B. (2002) *The future of the capitalist state*. Cambridge: Polity Press.

Kemp, D. and Dowling, P. (2000) Towards sustainable perennial pastures, *Australian Journal of Experimental Agriculture*, 40, 125–132.

Kenney, M., Lobao, L., Curry, J. and Goe, R.(1991) Agriculture in US Fordism: the integration of the productive consumer, In Friedland, W., Busch, L., Buttel, F. and Rudy, A.(eds), *Towards a new political economy of agriculture*. Boulder, Colorado: Westview Press, pp. 173–188.

Lawrence, G. (1987) *Capitalism and the countryside*. Sydney: Pluto Press.

Lawrence, G. (1989) The rural crisis down under: Australia's declining fortunes in the global farm economy, In Goodman, D. and Redclift, M. (eds), *The international farm crisis*. Basingstoke: Macmillan, pp. 234–274.

Lawrence, G. (1990) Agricultural restructuring and rural social change in Australia, In Marsden, T., Lowe, P. and Whatmore, S. (eds), *Rural restructuring: global processes and their responses*. London: David Fulton, pp. 101–128.

Lawrence, G., Vanclay, F. and Furze, B.(eds) (1992) *Agriculture, environment and society: contemporary issues for Australia.* Melbourne: Macmillan.

Lawrence, G. and Gray, I. (2000) The myths of modern agriculture: Australian rural production in the 21st century, In Pritchard, B. and McManus, P. (eds), *Land of discontent: the dynamics of change in rural and regional Australia.* Sydney: University of New South Wales Press, pp. 33–51.

Lobao, L. and Meyer, K. (2001) The great agricultural transition: crisis, change, and social consequences of twentieth century US farming, *Annual Review of Sociology,* 27, 103–124.

Lockie, S. and Bourke, L. (eds) (2001) *Rurality bites: the social and environmental transformation of rural Australia.* Sydney: Pluto Press.

Marsden, T. (2003) *The condition of rural sustainability.* Assen: Van Gorcum.

Marsden, T. and Sonnino, R. (2005) Rural development and agri-food governance in Europe: tracing the development of alternatives, In Higgins, V. and Lawrence, G. (eds) *Agricultural governance: globalization and the new politics of regulation.* London: Routledge, pp. 50–68.

Martin, P. (1997) The constitution of power in Landcare: a post-structuralist perspective with modernist undertones, In Lockie, S. (ed.) *Critical Landcare,* Wagga Wagga , NSW: Charles Sturt University, pp. 45–56.

Martin, P. and Ritchie, H. (1999) Logics of participation: rural environmental governance under neo-liberalism in Australia, *Environmental Politics,* 8, 117–135.

McCarthy, J. and Prudham, S. (2004) Neo-liberal nature and the nature of neo-liberalism, *Geoforum,* 35, 275–283.

McMichael, P. (2003) The impact of global economic practices on American farming, In Brown, D. and Swanson, L. (eds), *Challenges for rural America in the twenty-first century.* University Park Pennsylvania: Pennsylvania State University Press, pp. 375–384.

McMichael, P. and Lawrence, G. (2001) Globalising agriculture: structures of constraint for Australian farming, In Lockie, S. and Bourke, L. (eds) *Rurality bites: the social and environmental transformation of rural Australia.* Sydney: Pluto Press, pp. 153-164.

Molnar, J. (2005) Rewarding the biggest, encouraging the rest: conservation security and environmental stewardship, unpublished paper presented to the RC 40 Mini-conference: An agriculture without subsidies? Visioning the challenges of a market driven agri-food system, Keszthely, Hungary.

Morgan, K., Marsden, T. and Murdoch, J. (2006) *Worlds of Food: place, power and provenance in the food chain.* Oxford: Oxford University Press.

Moyer, W. (2004) Agricultural policy change in the EU and the US, unpublished paper presented to the XI World Congress of Rural Sociology, Trondheim, Norway.

Orden, D. (2002) The farm policy reform process, In Moss, C., Rausser, G., Schmit, A., Taylor, T. and Zilberman, D.(eds), *Agricultural Globalization, Trade and the Environment.* Dordrecht, Kluwer, pp. 5–20.

Orden, D. (2003) *US agricultural policy: the 2002 Farm Bill and the WTO Doha Round proposal.* Washington DC: International Food Policy Research Institute.

Orden, D. *et al.* (1999) *Policy reform in American agriculture.* Chicago: University of Chicago Press.

Petit, M. (2002) The new US Farm Bill: lessons from a complete ideological turnaround, *Eurochoices*, Winter.

Potter, C. and Tilzey, M. (2005) Agricultural policy discourses in the European post-Fordist transition: neo-liberalism, neo-mercantilism and multifunctionality, *Progress in Human Geography*, 29, 581–601.

Potter, C. and Tilzey, M. (2007) Agricultural multifunctionality, environmental sustainability and the WTO: resistance or accommodation to the neo-liberal project for agriculture? *Geoforum*, 38, 1290–1303.

Pritchard, B. (2000) Negotiating the two-edged sword of agricultural trade liberalization: trade policy and its protectionist discontents, In Pritchard, B. and McManus, P. (eds) *Land of discontent: the dynamics of change in rural and regional Australia.* Sydney: University of New South Wales Press, pp. 92–104.

Pritchard, B. (2005) Implementing and maintaining neoliberal agriculture in Australia. Part 1: Constructing neoliberalism as a vision for agricultural policy, *International Journal of Sociology of Agriculture and Food*, 13, 1–12.

Prober, S. and Thiele, K. (1995) Conservation of the Grassy Whitebox Woodlands: relative contributions of size and disturbance to floristic composition and diversity of remnants, *Australian Journal of Botany*, 43, 349–366.

Ritchie, M. and Ristau, K. (1986) US farm policy: the 1987 debate, *World Policy Journal*, 4, 32–40.

Sustainable Agriculture Coalition (2003) Re: advanced notice of proposed rule-making for the Conservation Security Program, letter to Natural Resources Conservation Service, USDA, March 12, 2003.

Tilzey, M. (1998) *Sustainable development and agriculture.* Peterborough: English Nature Research Report 278, English Nature.

Tilzey, M. (2000) Natural Areas, the whole countryside approach and sustainable agriculture, *Land Use Policy*, 17, 279–294.

Tilzey, M. (2002) Conservation and sustainability, In Bowler, I.R., Bryant, C.R. and Cocklin, C. (eds), *The sustainability of rural systems: geographical interpretations.* Dordrecht: Kluwer, pp. 147–168.

Tilzey, M. (2005a) Agriculture, trade and multifunctionality: the political ecological contexts for 'post-productivist' initiatives in Australia and the EU, unpublished paper presented to the National Europe Centre Public Seminar Series, Australian National University, Canberra.

Tilzey, M. (2005b) Changing agri-rural governance in the European post-Fordist transition: the challenge for sustainability in the uplands, unpublished paper presented to the Bioscene Conference, Ioannina, Greece.

Tilzey, M. (2006) Neo-liberalism, the WTO and new modes of agri-environmental governance in the European Union, the USA and Australia, *International Journal of Sociology of Agriculture and Food*, 14 (1), 1–28.

Tilzey, M. and Potter, C. (2007) Neo-liberalism, neo-mercantilism, and multi-functionality: contested political discourses in European post-Fordist rural governance, In Cheshire, L., Higgins, V. and Lawrence, G. (eds) *International perspectives on rural governance: new power relations in rural economies and societies.* London: Routledge.

Van Apeldoorn, B. (2002) *Transnational capitalism and the struggle over European integration.* London: Routledge.

Van der Pijl, K. (1998) *Transnational classes and international relations.* London: Routledge.

Walford, N.S. (2003) Productivism is allegedly dead, long live productivism: evidence of continued productivist attitudes and decision-making in south-east England, *Journal of Rural Studies*, 19, 491–502.

Williams, C. (2004) *Old land, new landscapes.* Melbourne: University of Melbourne Press.

Wilson, G.A. (2001) From productivism to post-productivism … and back again? Exploring the (un)changed natural and mental landscapes of European agriculture, *Transactions of the Institute of British Geographers*, new series, 26, 77–102.

Wood, E. (2002) The question of market dependence, *Journal of Agrarian Change*, 2 (1), 50–87.

Wood, E. (2005) *Empire of capital.* London: Verso.

Zinn, J. (1999) *Conservation spending in agriculture: trends and implications.* Washington DC: USDA, CRS Report RL 30331 October.

PART 2
Sustainable Agriculture

Chapter 3

Constructing Sustainability Through Reconnection: The Case of 'Alternative' Food Networks

Rosie Cox, Moya Kneafsey, Laura Venn, Lewis Holloway,
Elizabeth Dowler and Helena Tuomainen

Introduction

This chapter draws on empirical research with five 'alternative' food schemes that allow consumers to access food outside the conventional retail complex. Each of the schemes involves direct contact between producers and consumers and with this the potential for (re)connection of consumers with the people and places that produce their food. The chapter explores the significance of notions and practices of sustainability for consumers involved in these schemes, and suggests ways in which the reconnection that some schemes allow are related to wider constructs of sustainability. We argue that the consumers/members[1] we talked to rarely spoke in terms of 'sustainability', but that a little probing revealed motivations that directly related to issues of social, economic and environmental sustainability. Furthermore, for some consumers, membership of such schemes had made them consider their consumption practices in some depth and had been a trigger for constructing more sustainable practices more broadly in their lives. It appears, therefore, that some 'alternative' food schemes can be powerful and attractive forces in supporting sustainable consumption and perhaps more sustainable lifestyles in both rural and urban areas. However, there are barriers to accessing such networks that currently limit their potential reach.

The chapter begins by briefly introducing 'alternative' food networks (AFN) and relating their emergence to moves towards greater sustainability in food supply systems. In particular, the ways in which social, economic and environmental sustainability intertwine are highlighted. We then go on to introduce our methods and give a brief description of our case study AFN projects. The bulk of the chapter reflects on findings from workshops and interviews with participants in these schemes. We discuss environmental, economic and social sustainability in turn and demonstrate the ways that scheme participants move from thinking about one of these areas into understanding their inter-relationship and perhaps altering their behaviour as a result.

What are 'Alternative' Food Networks?

'Non-conventional' or 'alternative' food networks can be conceptualised in relation to a globalised 'conventional' food system, where food is transformed and transported across international boundaries within complex and elongated industrial food chains that remove foods from their origins, rendering food production invisible and primary food products unrecognisable to end consumers (Fischler, 1988; Holloway and Kneafsey, 2004; Collet and Mormont, 2003; Cook and Crang, 1996). Conversely, food sourced from 'alternative' food networks, is largely considered to have been produced, processed and distributed, and consumed within a given region or locality without the need for elongated, multi-actor food chains. As such, non-conventional food networks allegedly resist and distance themselves from the omnipresent industrialised mode of food production and consumption, by reconnecting producers, consumers and their food.

The interest in non-conventional food networks, or 'alternative' (agri-) food initiatives as they are sometimes referred to, predominantly emerged in the late 1980s as food initiatives, such as local/community food projects, began to appear, with specific agendas to negotiate closer relationships between food producers and food consumers. At the same time concern about intensive food production underpinned growth in organic production and other practices focused on environmental sustainability. There are now many variants of AFN, some with explicit agendas for reconnection and/or sustainability, but others without. Various writers have identified AFN as including: Community Supported Agriculture (CSA) schemes, cooperatives, urban gardens, farmers' markets, community land trusts, food policy councils, direct marketing, value-added marketing, school farms and sustainable agriculture organisations (DeLind and Ferguson, 1999; Clancy, 1997; Feenstra, 2002; Grey, 2000; Lacy, 2000; Pretty, 1998). However, precise delineation is problematic as the literature contains a wide variety of definitions and often-conflicting constructions of what constitutes an AFN. Indeed, almost any attempt to supply individuals, groups, organisations or institutions via distribution channels that are detached from the centralised and corporately-owned supply networks of dominant conglomerates can be portrayed as exemplars of 'alternative' food networks. Thus, the AFN terminology frequently describes diverse agri-food networks that may include varying forms of food production, food commodities or food relationships between producer and consumer that outwardly appear to differ from the standardised, industrialised mode of food supply (Murdoch *et al*, 2000).

One way of thinking through the relationship between AFN and sustainability draws on Goodman's (2003) proposition that there have been two more or less distinctive bodies of literature on 'alternative' food networks emerging from the different approaches and perceived priorities associated with European and North American scholars. The term 'alternative' has had different nuances in these different literatures and offers different possibilities of sustainable practices. On the one ('European') hand, the 'alternative' has generally been regarded as that which can fit into the interstices, or around the margins, of a 'conventional' industrial food supply system as a means for small businesses to survive in an aggressively competitive market. On the other ('North American') hand, the 'alternative' is regarded in more radical terms as something oppositional to industrial food supply and relates both to a wider sense of protest, and to attempts to establish different modes of exchange between food producers and consumers.

In the first ('European') case, according to Goodman (2003), 'alternative' has become synonymous with a series of other terms, such as 'local', 'reconnection' and 'quality' (see also Whatmore *et al*, 2003). Food systems which exhibit these qualities (in themselves difficult to define) have been depicted as a response to rural economic 'crisis', the need for particular trajectories of rural development, and as a response to consumer anxieties about food (Goodman 2003; 2004; Stassart and Whatmore, 2003; Whatmore *et al.*, 2003). For Goodman (2003), in this case, 'alternative' food networks represent niche development opportunities within an existing mass food supply system, which is not threatened by their emergence and which in fact may absorb and benefit from some of the associated ideas. Within this literature Gilg and Battershill (1998) discuss the 'alternative' in terms of food 'quality', suggesting that 'quality' food systems might present an 'alternative' to the industrialised food sector as part of rural development strategies. Similarly, Sage (2003) describes the 'alternative' in terms of 'good food' which might be associated with a re-embedding of food supply systems in localised social contexts, implying too a high degree of social connectivity associated with food production and consumption. Here, 'alternative' food networks are those which are seen as being 're-embedded' in their local social contexts.

In the second ('North American') case, Goodman (2003) argues, 'alternative' has become associated with a more explicitly politicised discourse of oppositional activism (Allen *et al*, 2003). Alongside a juxtaposition of 'alternative' with ideas of localness and quality which mirrors that of the 'European' literature, the heterogeneity of that which is covered by 'alternative' here is clearly acknowledged, although many authors seek to unite the various 'alternatives' in relation to particular politicised discourses. Grey (2000) recognises this in saying that the 'alternative' supply system "promotes a new consciousness about the sources and quality of food, an awareness that cannot be separated from the social and political dimensions of food production" (p. 147). Similarly, Allen *et al* argue that 'alternative' food networks share an agenda which focuses on creating food systems which represent attempts by some producers and some consumers to wrest control over food supply from actors in an industrialised mass food supply system. Thus, as they argue, these food networks "share a political agenda: to oppose the structures that coordinate and globalise the current food system and to create alternative systems of food production that are environmentally sustainable, economically viable and socially just" (2003: 61; see also Hendrickson and Heffernan, 2002). Discussion from this perspective implies that changing the way food is produced and consumed has the potential for engineering larger scale social change, engaging producers and consumers in a wider political struggle surrounding the sorts of social, economic and ecological relationships that people would like to exist (see, for example, Hassanein, 2003). This politicisation of food networks implies the involvement, as activists, of food producers and consumers in order to create 'alternative' food systems which display the characteristics they desire. These characteristics, such as 'trust', 'authenticity' and 'embeddedness', again resemble those suggested in the 'European' literature described above, although in this case they become allied to a more radical agenda for change. Thus, for example, Hendrickson and Heffernan (2002) argue that 'alternative' food movements need to "organise where the dominant system is vulnerable – by making ecologically sound decisions, by relying on time and management rather than capital, and by building authentic trusting relationships that are embedded in community" (p. 361).

These two models, therefore, offer us two different versions of sustainability – one focused on an idea of sustainable rural economies/communities, which simultaneously emphasises enterprise, small businesses, conserving particular 'traditional' rural ways of life, landscapes etc, while the second is a more 'radical', politicised sustainability which is more progressive in terms of seeking social and environmental justice.

In both literatures great emphasis has been put on the needs and practices of producers. Indeed, Goodman's (2003) distinction relies on literature which has tended to concentrate on producers and production, to the relative neglect of consumers and consumption. There thus remain many questions and unknowns about the role of consumers, their objectives and desires and their potential as contributors to more sustainable food systems. In our research we have worked with a small number of AFN that offer the opportunity for direct contact between producers and consumers and with such contact the possibility of 'reconnection'. Rather than looking just at the needs of food producers and their strategies for maintaining a livelihood, we are focusing on what consumers want, how reconnected they want to be and what consumers want from food producers.

Sustainability and Reconnection

As Robinson (2004) suggests, the notion of sustainability has arrived only relatively recently into thinking about agricultural development, stemming from international negotiations concerning environmental damage and protection at the global scale (e.g. the Rio 'Earth Summit', 1992, and the incorporation of environmental concerns into thinking about economic development evidenced in the EU's 5th Environmental Action Programme, 'Towards sustainability', running from 1993-2000) (Cobb *et al*, 1999; Bowler, 1999, but see Beus and Dunlap, 1990). Understandings of sustain-ability have tended to coalesce around environmental issues, whether at global or smaller scales. Yet the notion evidently extends to thinking about social and economic conditions, and more holistic conceptualisations have tried to show how the environmental and other dimensions of sustainability are inter-related. In the context of agriculture, Troughton (1993; 1997 cited in Robinson, 2004), for example, presents a nested hierarchy of agricultural sustainability, beginning at the scale of the field ('agronomic sustainability'), and running through the scales of the farm business ('micro-economic sustainability'), the local community ('social sustainability') and the national economy ('macro-economic sustainability'), ending at the scale of the global environment ('ecological sustainability').

By also thinking about rural sustainability through food consumption practices other aspects of the inter-relationship between society, environment and economy can be brought into focus. As part of the wider emergence of 'ethical' consumption, where consumers articulate concerns for the social and ecological effects of their consumption by consuming, or not consuming, particular goods or services, sustainable consumption practices have been described as including behaviours such as the selection of 'green' products (phosphate-free detergents, dolphin-friendly tuna etc), the rejection of wasteful packaging on products, and the rejection of products containing certain harmful additives, such as CFC propellants; all actions that relate directly to environmental protection (Cairnross, 1991). However, our research suggests that some participants in AFN are able to make broader and more nuanced connections between their consumption practices and the lives and livelihoods of producers in rural areas and local communities.

The Case Studies

The findings drawn on in this chapter were gathered as part of a wider project titled 'Reconnecting consumers, food and producers: exploring "alternative" networks'.[2] The project focussed on six case study AFNs that were selected after an initial scoping exercise conducted in the first six months of the project revealed a wealth of non-conventional food networks that in various ways aimed to reconnect producers and consumers (Venn *et al.*, 2006). The six case studies selected showed the diversity of schemes available and they involve varying degrees of consumer engagement and involvement in food production. All of the schemes involve some kind of direct contact between producers and consumers and all would be thought of (and some specifically aim to be) more sustainable than traditional intensive agriculture. This chapter draws on workshops and interviews with participants in five of the schemes, all located in various parts of Britain. They are:

- Waterland Organics, a box scheme based near Cambridge that provides weekly boxes of vegetables to 80 households. They are also members of Eostre Organics, a producer cooperative based in East Anglia;
- EarthShare, a CSA project based near Forres in NE Scotland which again provides weekly and fortnightly vegetables boxes to around 160 households. In addition to paying a subscription members of this scheme also have to do three workshifts a year;
- Farrington's Farm Shop, located in Farrington Gurney near Bristol. A well-established farm shop with currently around 8000 customers registered on their loyalty card scheme;
- Moorland Farm, which raises Aberdeen Angus-sired cattle and sells their meat directly to the public through their own farm shop and various farmers' markets;
- Salop Drive Market Garden, a food cooperative based in Sandwell in the urban West Midlands. They provide approximately 50 households with bags of vegetables during the summer months.

The sixth case study, 'Adopt-a-Sheep', is based in central Italy and has consumers all over the world. Findings from this case study are not included here as we have used different methods to contact participants in that scheme (see Holloway, 2002; Holloway and Kneafsey, 2004; and Holloway *et al.*, 2006).

The first stage of research involved interviews with producers/managers of schemes and workshops with participants in the schemes. In these workshops we asked the consumers/members of schemes what they liked about the scheme and why they got involved. We also asked them about how they thought the scheme contributed to sustainability. In the second stage of research we carried out in-depth interviews with participants and probed these issues further. We have gathered detailed data on their food shopping habits and their reasons for selecting the foods that they do. In the next section we present some of our findings in detail and discuss how consumers conceptualised schemes and justified participation (in environmental, economic and social terms); how participation had impacted on wider consumption practices; and the challenges to being a conscientious consumer.

Consumers' Conceptualisation of AFN

In many cases consumers' participation in AFN was explained in terms of personal reasons and desires, such as the pleasure afforded by the 'surprise' contents of an organic box or the convenience of delivery. However, many consumers also voiced their appreciation of issues which relate broadly to sustainability, such as perceived social and economic benefits to local communities and in some cases the environmental gains of purchasing food through distinct supply chains. Discussions with consumers about their participation revealed that many took an interest in the impacts of their actions. Some interviewees were well informed and knowledgeable, others articulated much vaguer notions about what was 'good' or 'bad' but still demonstrated a desire to care for people and environments that might have been thought of as remote from them.

Environmental Justifications for Participation in AFN

Sustainability is often associated with the environment, yet our workshop participants talked about this attribute far less than others. No-one said they joined AFN because they saw it as a sustainable option. However, some did argue that by supporting less intensive farming, whether this be farming using 'organic' methods or small-scale production with just one farmer growing just enough to sell at market, there were potential benefits for the wider environment, in particular wildlife and biodiversity. As one participant put it:

> [I]it isn't the food, it isn't the food so much, that's partly it, but it's really because you can't say that the present generation owns the soil ... I mean, we feel this very deeply indeed, you cannot hand over to your incoming people, your incoming generations, soil that is so desperately polluted and inert, you know, it's lost it's life, and it's awfully true to say this because pollution and inert, it's lifeless, and you've only got to look at the fields that have been conventionally treated. (EarthShare)

Furthermore, several consumers were familiar with the notion of food miles and the contribution they were making to reduce pollution by sourcing their food locally. As one interviewee put it:

> I think it's wrong that um all the ... and you hear all about the global warming and all the ... and you think of all the fuel that ... that they use to fly into all them ... it's just ridiculous and that's another thing, you see, the big ... the big ... the big um Tescos, the Morrisons and Sainsbury's they've huge great lorries travelling from one end of the country to the other. I mean why should they have to do that? (Farringtons)

For another, the links were quickly made between her consumption practices and support for social and environmental sustainability in the local area and around the world.

> INTERVIEWER: Could you explain to me a little bit more about why it is important to you to buy local?
>
> RESPONDENT: Um, well it's environment really, isn't it, resources and transporting things across the world and yes, ... I just think you should support your local community, I think. And it's also things like if you buy South American beef and how much rain forest has been cut down to rear them. (Farringtons)

These comments illustrate that while 'sustainability' was not a concept that had resonance with the consumers interviewed, care for the environment was important to them in a number of ways that resonate with both of Goodman's (2003) 'European' and 'North American' constructions of AFN. However, it was in discussions of economic and social relationships that the true potential of AFN to contribute to a notion of more sustainable food systems revealed itself, and it is to these aspects that we now turn.

Economic Justifications for Participation

Like many consumers, the participants in AFN that we talked to were interested in the cost of their food. Household incomes of participants using the schemes varied widely both within and between schemes and the majority were by no means the 'trendy middle-classes' who are often thought to populate AFN. The Salop Drive project was set up specifically to provide fresh produce to local people on low incomes and the other schemes providing organic fruit and vegetables were confident that their produce was cheaper than supermarket alternatives. The producers and participants in these schemes were, therefore, price conscious rather than seeking alternative sources at any price. In fact the most frequently mentioned economic justifications for using a specific scheme were that it was perceived to be good value or inexpensive compared to more mainstream sources.

However, where premiums were experienced, consumers reported that they were willing to pay slightly more because they were aware that the money stayed in the local community and it was evidently funding local jobs and employment, which were seen as positive and legitimate reasons for paying slightly higher prices. Some consumers made links between the price of food and environmental sustainability and, as the quote below illustrates, some made links between cost, environment and power within the food chain more generally:

> [T]he supermarkets are doing you no favours by selling cheap food because cheap in terms of food means bad. Yeah, something is wrong with it. You know I said it's either going out of date or it's that the farmer at the end is going to commit suicide because he can't feed his own family let alone the rest of you and I said something is wrong. And you know it's like the chickens and the fact that you can buy two chickens for a fiver or something, they are, they are beyond belief the way these chickens are brought in from Indonesia. They all cost 10p, they're cryogenically frozen in some kind of suspension, so that when they get here they're unfrozen by a chemical process and then they say they're fresh. Because the fact they haven't been technically frozen as we would understand it, they've got a big union jack on them which in minute writing, probably in a foreign language, says 'packaged in the UK', and they sell them for £2.99 as fresh, and can be frozen, which they technically can. They're pumped full of antibiotics, so they reckon in 20 years we will have no immunity anymore to antibiotics. It just freaks you out. (Moorlands)

There is some evidence from our research that rather than consumers being simply divided into those who want cheap food (at any cost to the environment) and those who want quality food (at any cost to themselves), the situation is in fact much fuzzier. Consumers are also capable of wanting good value without detriment to others or to the natural environment and may make well-informed and nuanced decisions about how to achieve this.

Consumers' Social Justifications for Participation

The social nature of transactions within the schemes featured in all workshop discussions and it became very clear as we spoke to people that these consumers valued the human contact that the schemes brought (although there were clear variations between schemes as well), and that this human contact differed significantly from human contact that they experienced at other food outlets. Our workshop participants were eager to impress on us that these relationships were about more than just the purchase of food. They represented new friendships, sharing information, a sense of community, pride and belonging. And it was being able to talk to the person who had actually produced the food that mattered most. Hence, socially these networks appear to possess very desirable attributes that tie people together and link them to wider communities.

Colin Sage (2003) has written about the relations of 'regard' that can develop within AFN, whereby consumers relate to producers with empathy and care. Such relationships offer the potential for a new logic within the food production system whereby producers and consumers can be allies striving together to create the conditions within which food production can benefit both groups. We found evidence for this type of friendship and concern between producers and consumers and within groups of consumers, particularly in those schemes that necessitated face-to-face and regular contact. For example, one participant from the Waterland scheme acknowledged that whilst personal contact with the grower was not an initial motivating factor for joining, she now appreciated that contact and would be less inclined to move to another scheme after having chatted to the producer on the doorstep about the produce and other topics, such as raising young children, when the box was delivered each week. As she explains:

> I think it was, yeah, it was more a secondary benefit [personal relationship with producer], now certainly it would mean that I wouldn't switch that easily now from [Waterland] to somewhere else.

Even those consumers shopping in the less intimate space of a farmers' market felt loyalty and relations of obligation developing when they were regular customers of the Moorland stall:

> [I]t's one of those things I think we were talking about before about the kind of … the loyalty to … you actually feel kind of a little bit guilty, walking past them in the morning, "They're such nice people. Oh, I should go and buy something from them"… I've said hello to them a number of times, um but um again I, you know, they were very friendly, very welcoming, and um I kind of got this impression you know their heart was really in it.

> I don't quite know what it is, I think it's that they, I don't know Moorland's seem as if … I wouldn't say they seem like they need your business but they, they seem … that's not what I mean, they seem like they really, are really keen on what they're doing. The other people seem to be more of a butcher, you know. I think they have all kinds of things and whether it's all their own produce, I don't know, but it doesn't come across as being that. Moorland's seem to be, you know, this is ours and it's, you know, it's like me selling something I've made. You know you feel that there … it really is their own produce, which it is.

It seems that personal relations between producer and consumer as well as being a means of communication and helping consumers to feel involved with the scheme, also help to retain consumers and support the continuation of AFN. Building up rapport with a 'real person' who represents the scheme can strengthen feelings of loyalty for some.

The social aspects of involvement in these schemes went beyond relationships with individuals and, for some participants, encompassed much broader notions of community, education and support for a way of life they believed in. As one EarthShare subscriber put it when describing doing the work shifts:

> It's a real community. It's lovely to meet other people and be involved in the scheme. I think that's one of the reasons why we know so many people because you know there are obviously other people that are working whenever you go along.

It was in these discussions that the greatest degree of 'reconnection' between consumers, producers and their food was expressed and, arguably, the possibilities for sustainable relationships in the food chain were revealed. The topics that consumers touched on in these discussions included: knowing the source of the food and those that grow it; sharing the risk with the farmer; bringing the countryside into the town; feeling part of the food chain; educating other adults and children; a sense of self-sufficiency in the community and eating off the same land as you live on. Often these philosophical, or ethical, goals were expressed in terms of what was good for others – such as the community and producer – as well as in terms of meeting the consumers' need to access appropriate food. As one commented about going to the Bristol Farmers' Market rather than to a supermarket:

> I hated doing it [going to a supermarket], especially with a small child in tow. I just thought 'Why am I spending 70 odd pounds a week here when I could be going and getting something locally that's only round the corner from me.' I'm at work on a Wednesday. So it's like five minutes' walk for me. I can go there; I can buy really nice fruit and veg, nice meat, and um and it's from the local community I feel rather than it's from a great big supermarket.

Another suggested that the fulfilment that comes from supporting a local business could help overcome the problem of the extra effort needed to prepare fresh produce:

> RESPONDENT: I mean, some evenings, yes, it can be a bit of a pain you know. You're home from work late, you're tired, but, if you are making a soup and you know the vegetables actually taste like vegetables.
>
> INTERVIEWER: Do you like the fact then? I mean, is local important to you?
>
> RESPONDENT: Yes, very, yeah.
>
> INTERVIEWER: So is it the dirt gives it some credibility that it's just come from …?
>
> RESPONDENT: Not so much that. It's about supporting your local community.

And an EarthShare subscriber made links between being in the scheme, a national 'food culture' and support for the local community.

> But it's the kind of whole ethos about the composting, you know, trying to sort of waste as little as possible, you know, feel more connected with the earth, more appreciative of the, the food that we're eating and, you know, getting back to some basic values that I think, you know, as a society we've lost and, you know, I do believe that, well maybe I would say Scotland, but England as well. In the UK we've kind of lost that food culture and this was, you know, now because I'm here, wanting to support a local community project and, you know, as best I can.

Despite rarely using the language of 'sustainability' many of our interviewees discussed desires and motivations for participating in AFN that displayed a real commitment to improving the lives of others and protecting both the local and wider environment. While many of the interviewees talked in terms of personal relationships, protecting the local economy, benefiting the local community and returning to 'traditional' practices and values, all themes that are reflected in the European AFN literature, our interviewees also moved easily between these ideas to broader notions of sustainability and social justice, supporting Grey's (2000) view that consciousness of the source and quality of food promotes an awareness of the social and political dimensions of food production.

'Alternative' Food Consumption and Sustainable Lifestyles

Many of the participants in AFN that we have interviewed appear to engage with food consumption thoughtfully and to pursue sophisticated strategies when accessing their food that could be a contribution to sustainability in the broadest terms. We explored whether these consumers' sustainable consumption in relation to a particular scheme was part of more generally sustainable behaviour and whether participation in an AFN scheme had made them change their actions in other areas.

Most commonly participants had thought about other areas of their food consumption. Several consumers spoke of their searches for less packaging on goods, and more buying of prepared rather than processed foods. A few commented on their new-found interest and desire to buy from other local sources, so we had people from EarthShare in Inverness saying that they now made sure they bought all their meat and fish from local butchers and fishmongers. A Farringtons' customer reported no longer using the supermarkets for anything other than general household goods and store cupboard goods. A successful AFN scheme appears to be a powerful tool for helping people think about their consumption, as these comments from two EarthShare participants show:

> I mean I was already kind of, you know, questioning things and … and looking at, you know, sort of Friends of the Earth member, trying to do kind of things that I can. Um, this [joining EarthShare] course kinda really opened it up probably too far, you know, at the early stages, because you then felt that you're swimming in it.

> But I still think starting with the eggs is like starting with the box because I think eventually it will … I mean one day somebody will think well you know we have all these free range eggs. Maybe I'll try a free range chicken, so again it is this gradual transition.

In addition to changes to where people shop, numerous participants reported that participation had also meant that their diets had positively changed as they were eating more vegetables and unusual items which previously they may not have bought. Consumers also spoke of eating more of what is in season. For some this could mean changes in shopping and cooking habits. For those with boxes it meant looking in their boxes, seeing what needed eating and then organising a meal. This was echoed by one of the farmers' market shoppers who now goes out without a shopping list and just buys what is available. These are significant departures from deciding on a recipe first and then going and searching for the ingredients irrespective of season.

Perhaps most interestingly we found that, for some people, participation in a scheme could influence their behaviour far beyond their food shopping. We found people who had become enthusiastic composters and recyclers. As one participant explained, thinking about the land where your food is grown can provide quite a direct route to thinking about your ecological impact more generally:

[A] lot of people join because of a feeling, an organic belief, it's right, but they can't quite think about the soil. They think primarily, which is good, about their tummies and the fact that they pour poison down their children's throats. That's fine, that's excellent, and then perhaps the primary ... that's the initial motivation, but after a little bit they begin to realise this grows in soil and if we treat soil as it's usually treated what on earth are we doing? And not only to soil but the watercourse, you see, it drains into the water, and it's ... we're very conscious of it up here. All that water goes into the bay and goes out into the Moray Firth. (EarthShare)

Another participant's comments illustrated that food's closeness to the body can provide a way of thinking about what else we are consuming for our bodies:

Then you sort of think, 'ok I've got these vegetables now and this is really working for me'. And then you maybe sat there and think oh well maybe I should be thinking about what I put in ... If I'm putting that in my body, what am I putting on my body? So then you think, 'oh, well actually I'd quite like to live the simpler lifestyle' and it just gets you thinking. (EarthShare)

Therefore food consumption might be a particularly good place for people to begin to think about their consumption practices in general because of its very direct and visible connections with the consumer's own body and the natural environment.

Conclusions: The Challenges to Being a Conscientious Consumer

Being a conscientious and sustainable consumer is not plain sailing. We uncovered constraints, contradictions and complications associated with using these alternative food networks, and some factors that may dissuade people from consuming responsibly or pursuing ethical motivations. In almost all discussions the benefits listed were countered by difficulties and barriers that make food buying the complicated and conditional practice that we all know it to be, which perhaps goes some way to explaining why reconnection is such a difficult thing to achieve.

From peoples' own experiences participation in these types of networks is not always easy and, whatever their motivations are, consumers come up against the practicalities of accessing food this way. The main limitations that came out from all our workshops were:

- Limitations of eating seasonally – produce availability, consistency and choice.
- Labour intensity – time and effort required to travel around different shops.
- The need to be organised – some sources are only available at specific times.
- Ties associated with feelings of loyalty.
- Accessibility – geographically and financially.

Hence being conscientious has costs as well as rewards. We asked our workshop participants why they thought other people did not get involved with these schemes, and, on top of the complications listed above, they identified a number of other reasons why people are unwilling and/or unable to participate in such schemes.

All of these potentially act as barriers to reconnecting producers with consumers within more sustainable food supply systems. First of all, people may be unaware of the options available to them, whereas those that do know of 'alternatives' may not understand, appreciate or value the existence of these types of food networks, perhaps perceiving them to be expensive or unsure of what they offer. Secondly, for those that do understand the principles behind these 'alternatives', participants acknowledged that people may fear change to their existing routines and as a result make excuses, for example claiming that they are too busy. Interviewees who use these schemes also reported that people who are inquisitive enough to ask them about their own participation often hold the perception that these things are in some way a compromise in quality or that the organisation of the scheme demands compromises – not choosing your own vegetables in a box scheme, being tied to a particular time etc. Finally, and very importantly, a significant barrier is people's access to food groups and local shops. Not everyone is fortunate enough to have outlets and schemes such as these on their doorstep.

Our interviewees showed us that while some participants in AFN were 'accidental' or not looking for dramatic alternatives to the mainstream, there are many other consumers who are informed, careful and committed, including some who are acutely aware of the principles behind these AFN and their contribution to a more sustainable mode of consumption. In many cases these people prided themselves on finding and supporting alternatives. While few of our participants would have described themselves as radical or characterised their food buying as 'political', our interviews uncovered a situation where many consumers were deeply committed to principles and modes of behaviour that are at odds with mainstream food production and perhaps in conflict with the thrust of globalised capitalism. And, for some, thinking about food had opened the door to questioning their lifestyles, and their social, economic and ecological effects, much more broadly.

Notes

1 We have struggled throughout the study to know how to refer to both those growing and supplying food and those receiving it. The terms 'producer' and 'manager' are used in this chapter for people operating schemes and 'participant', 'consumer' or 'member' for those accessing food through them. None of these terms is quite right for all the schemes but they are intended to convey the *relationships* involved between individuals, and between individuals and a scheme.

2 This research was funded under the ESRC-AHRC Cultures of Consumption Programme. Project reference no. RES-143-25-0005.

References

Allen, P., Fitzsimmons, M., Goodman, M. and Warner, K. (2003) Shifting plates in the agrifood landscape: the tectonics of alternative agrifood initiatives in California. *Journal of Rural Studies* 19: 61–75.

Beus, C. and Dunlap, R. (1990) Conventional versus alternative agriculture: the paradigmatic roots of the debate. *Rural Sociology*, 55: 590–616.

Bowler, I. (1999) Developing sustainable agriculture. *Geography*, 87: 205–212.

Cairncross, F. (1991) *Costing the Earth*. Random House Business Books, London

Clancy, K. (1997) Reconnecting farmers and citizens in the food system. In: Lockeretz, W. (ed.) *Visions of American agriculture.* Iowa State University Pres: Ames), pp. 47–57.

Cobb, D., Dolman, P. and O'Riordan, T. (1999) Interpretations of sustainable agriculture in the UK. *Progress in Human Geography*, 23: 209–235.

Collet, E. and Mormont, M. (2003) Managing pests, consumers and commitments: the case of apple growers and pear growers in the Lower Meuse region. *Environment and Planning A*, 35: 413–427.

Cook, I. and Crang, P. (1996) The world on a plate: culinary culture, displacement and geographical knowledges. *Journal of Material Culture* 1: 131–153.

DeLind, L. and Ferguson, A. (1999) Is this a women's movement? The relationship of gender to community supported agriculture in Michigan'. *Human Organization* 58: 190–200.

Feenstra, G. (2002) Creating space for sustainable food systems: lessons from the field. *Agriculture and Human Values*, 19: 99–106.

Fischler, C. (1988) Food, self and identity. *Social Science Information*, 27: 275–292.

Gilg, A. and Battershill, M. (1998) Quality farm food in Europe: a possible alternative to the industrialised food market and to current agri-environmental policies: lessons from France. *Food Policy*, 23: 25–40.

Goodman, D. (2003) Editorial: the quality 'turn' and alternative food practices: reflections and agenda. *Journal of Rural Studies*, 19: 1–7.

Goodman, D. (2004) Rural Europe redux? Reflections on alternative agro-food networks and paradigm change. *Sociologia Ruralis*, 44: 3–16.

Grey, M. (2000) The industrial food stream and its alternatives in the United States: an introduction. *Human Organization*, 59: 143–150.

Hassanein, N. (2003) Practicing food democracy: a pragmatic politics of transformation. *Journal of Rural Studies*, 19: 77–86.

Hendrickson, M. and Heffernan, W. (2002) Opening spaces through relocation: locating potential resistance in the weaknesses of the global food system. *Sociologia Ruralis*, 42: 347–369.

Holloway, L. (2002) Virtual vegetables and adopted sheep: ethical relations, authenticity and internet-mediated food production technologies. *Area*, 34: 70–81.

Holloway, L. and Kneafsey, M. (2004) Producing-consuming food: closeness, connectedness and rurality in four 'alternative' food networks. In Holloway, L. and Kneafsey, M. (eds) *Geographies of Rural Cultures and Societies.* London: Ashgate, pp. 257–277.

Holloway, L., Venn, L., Cox, R., Kneafsey, M., Dowler, E. and Tuomainen, H. (2006) Managing sustainable farmed landscape through 'alternative' food networks: a case study from Italy. *Geographical Journal*,172: 219–229.

Lacy, W. (2000) Empowering communities through public work, science and local food systems: revisiting democracy and globalisation. *Rural Sociology*, 65: 3–26.

Murdoch, J., Marsden, T.K., and Banks, J. (2000) Quality, nature and embeddedness: some theoretical considerations in the context of the food sector. *Economic Geography*, 76: 107–125.

Pretty, J. (1998) *The living land: agriculture, food and community regeneration in rural Europe.* London: Earthscan Publications.

Robinson, G.M. (2004) *Geographies of agriculture: globalisation, restructuring and sustainability.* Harlow: Pearson.

Sage, C. (2003) Social embeddedness and relations of regard: alternative 'good food' networks in south-west Ireland. *Journal of Rural Studies*, 19: 47–60.

Stassart, P. and Whatmore, S. (2003) Metabolising beef: food scares and the un/ remaking of Belgian beef. *Environment and Planning A*, 35: 449–462.

Venn, L., Kneafsey, M., Holloway, L., Cox, R., Dowler, E. and Tuomainen, H. (2006) Researching European 'alternative' food networks: some methodological considerations. *Area*, 38: 248–258.

Whatmore, S., Stassart, P. and Renting, H. (2003) Guest editorial: What's alternative about alternative food networks? *Environment and Planning A*, 35: 389–391.

Chapter 4

Farm Animals and Rural Sustainability

Nick Evans and Richard Yarwood

Livestock as Neglected Subjects

Farm livestock animals remain something of an enigma to the rural sustainability debate. A standardised view of livestock has emerged whereby farm animals are simply 'there' in the rural landscape (Evans and Yarwood, 1995). Recent concerns central to the assessment of agriculture in the UK as a sustainable practice, such as a shift towards organic farming systems (Lampkin, 1990), the (re)localisation of food (Rickets-Hein *et al.*, 2006) and the reassociation of agriculture with the production of biodiversity (Whitby, 1994), have done little to establish livestock as a focus for analysis. It seems that livestock are too often assumed to be an end point of a change in the conduct of agriculture rather than as active constituents in the project to create a better global environment. Take, for example, the case of organic farming where it is the initial decision by a farmer to 'go organic' that is viewed as a monumental leap along the path of sustainable development. General discussions about landscape impacts, the economic equation between a reduction of inputs versus a decline in productivity, gains in biodiversity and effects on food quality soon follow. In the melée of these discourses (see Goodman and DuPuis, 2002), questions are rarely raised about the consequences on livestock as a whole, let alone the impact on *livestock breeds* which has been central to our research over the last 12 years.

It could be countered that a focus on quality food during the 1990s drew attention to livestock breeds through the individuality of meat products, such as revealed by the promotion of Aberdeen Angus steaks as free from Bovine Spongiform Encephalopathy (BSE). Again, this situation is very much illustrative of the end point assumption, one that denies the efforts of those new animal geographers who attempt socially to redefine human-animal boundaries by radically moving animals away from their conventional status as observable natural objects at the margins of geographical enquiry (Philo and Wilbert, 2000). Rather, this chapter, using a discursive approach, attempts to reappoint livestock to their rightful place as essential threads in the tangled web of rural sustainability (Whatmore, 1997). Following brief reference to the ways in which livestock are commonly viewed, the links between species of farm animals, particular breeds and sustainability are explored. A detailed discussion of sustainability itself is beyond the scope of this chapter, but what is clear is that modern farming practices have dramatically reshaped the livestock mix within the UK agricultural sector to the point where major change seems inevitable given the new societal goals assigned to UK farming (see Policy Commission on the Future of Farming and Food, 2002). As illustrated in the subsequent section, policy directives up to now have been slow to respond and thin in content. As the importance of livestock breed becomes more widely appreciated, we conclude that future generations will observe animals that have a specific function, as yet virtually unknown, relating to the fulfilment of key biodiversity objectives (as encapsulated in Biodiversity Action Plans) delivered from different types of (sustainable) farming.

Viewing Livestock

For many members of the general public, livestock will not appear to have changed much during their lifetime. Sheep look pretty much alike and cows generally seem still to be the 'proper' black and white animals they vaguely remember as children. Perhaps the only observable effect, excepting the temporary wholesale removal of animals from the British landscape during the 2001 Foot and Mouth Disease (FMD) epidemic, is that there are less stock and more crops in the modern agrarian landscape. This can reasonably be accounted for by a shift in the balance of farm profitability away from stock-based enterprises towards crops and indoor, intensive livestock systems, accelerated through UK government policy mechanisms under the umbrella of the Europe Union's Common Agricultural Policy (CAP). Even with powerful policy signals, it could be argued, as Whatmore (2002: 97) has cited from Simmons' (1979: 11) book on '*Principles of Crop Improvement*', that:

> Probably, the total genetic change achieved by farmers over the millennia was far greater than that achieved by the last hundred or two years of more systematic-based effort.

If this is true of crops, grass and other vegetation used as agricultural 'forage', then it is unsurprising to consider change as subtle with more transient livestock. Those exposed fully to the phenomenon of livestock change over time have been restricted to membership of societal groups inhabiting the same domains as the animals themselves; the farmers, the societies representing particular breeds of livestock and specialist interest groups whose declared function is the conservation / preservation of the genetic identity of breeds (see Yarwood and Evans, 2006; Evans and Yarwood, 2000).

Some researchers have engaged with the issue of livestock change from genetic or historical perspectives (Bowman and Aindow, 1973; Walton, 1984). However, it is notable that agricultural geographers have largely treated livestock as units of production to be mapped over space; agricultural economists have engaged with livestock as subsets of various factors of production and animal scientists have treated livestock as a means to an end either in the unwavering scientific focus on raising food output or achieving more effective medical treatments for human illnesses. Only in more recent times have efforts been made to consider livestock in a more critical and culturally informed manner (Yarwood and Evans, 2000; Morris and Evans, 2004). Further, the growth of non-governmental environmental organisations that have done so much to raise public conservation awareness in the arena of countryside change has yet to exert an influence on the inclusion of livestock. The main 'pressure group' concerned with British livestock is the Rare Breeds Survival Trust (RBST) (see Evans and Yarwood, 2000). However, current membership stands only at around 8000 people compared with more than one million members of the

Royal Society for the Protection of Birds (RSPB) which is concerned with bird conservation or the 115,000 members of the 'Woodland Trust' (Haezewindt, 2003). Indeed, even the specialist charity 'Butterfly Conservation' has a membership of 11,500 demonstrating the lowly position of livestock within public imagination (see Holloway, 2004, for one discussion of how the agricultural sector presents itself to the public). Other than this, it is the societies associated with individual breeds that represent the core support for specific types of livestock. These have their roots in the age of nineteenth century agricultural improvement and ever since have been developing discourses to promote both the animals of their patronage and the status of their own organisation (Ritvo, 1987; Yarwood and Evans, 2006).

Unsustainable Livestock?

It is apparent from the preceding discussion that crucial questions relate to how livestock have changed and the extent to which they have become 'unsustainable' within modern farming systems. It is well-documented that the agricultural 'revolutions' of the eighteenth and nineteenth centuries began to quicken the pace of livestock change (Clutton-Brock, 1981). At first, this was largely a simple question of trial and error in breeding animals. One of the earliest and most influential practitioners was Robert Bakewell of Leicestershire. He broke with the practice of mixing animals of different breeds within the same field which had been based on the assumption that animals inherited their characteristics from the land. He isolated animals of similar type and was able to fix their characteristics, still seen in some breeds today such as the Leicester Longwool sheep. Although making a key contribution to what was to become scientific stockbreeding, his rudimentary understanding of genetics led to persistent in-breeding of animals, and thus to the eventual collapse of his stock enterprises.

As scientific methods became established by the Victorians, experimental livestock breeding became widespread amongst landed entrepreneurs. This reflected not only an interest in breeding animals capable of feeding the expanding urban industrial workforce, but also the currency of livestock, particularly large animals, as status symbols. For example, the Dorset Gold Tip pig was one culmination of breeding for size. The Gold Tip was so large that it was unable to move out of its pen and a forked stick had to be used to prop up its snout to avoid the animal suffocating within its own layers of fat. The first part of the twentieth century saw the application of Mendelian genetic principles to livestock production and some trans-national breeding effort, but without a serious undermining of breed diversity.

It was in the immediate post-war period that, at the broadest level, livestock began to undergo a rapid process of 'modernisation' as agriculture was assigned a trajectory based on an agro-industrial developmental model. Bowler (1985a) classified the structural dimensions of such agrarian change as intensification, concentration and specialisation, identifying responses and consequences. These are wide-ranging, cutting across all aspects of what is commonly termed 'productivist' agricultural practice (for a summary, see Robinson, 2004). On reflection, each has acted upon livestock in a specific manner, which Table 4.1 attempts to tease out, but to the same ultimate effects – a loss of breed diversity and increasingly dubious levels of sustainability.

Beyond such observations which apply to all livestock, it is necessary to examine more closely livestock animals by species and breed to come to terms with the subtlety of change previously noted. The next sections consider in turn how cattle, sheep and pigs may have deviated from the path of sustainability in their respective production systems.

Table 4.1   The modernisation and industrialisation of livestock

Structural dimension	Specific outcome for livestock
Intensification	• Higher stocking densities • A move to indoor production systems • A focus on 'productive' breeds (time and quantity of food product) at the expense of the less productive • An acceleration of genetic control through scientific breeding practices (eg. artificial insemination) • The use of 'champion' animals judged on productivity, narrowing the genetic range • Changing animal diets • Greater use of veterinary products
Concentration	• Fewer farm units with livestock • More livestock per farm unit • Stronger regional association with stock production • Greater dependence upon contract forms of meat and milk production, increasingly influenced by large processing businesses • Closure of small abattoirs • Fewer farmers with experience of livestock: erosion of local and traditional knowledges • Decision-making about livestock increasingly policy-led and reactionary
Specialisation	• A greater commitment to one type of livestock production • Reduced flexibility to mix livestock enterprises • A more economically ruthless focus on which types of livestock can exist on the farm • A loss of specialised labour • Increased disease risks • Greater incidences of stock movement and over longer distances • Lack of contact with like-minded breeders • Reduction in breed society members • Severing of the link between production and local consumption

Cattle

Successive influxes of peoples into the British Isles have left a rich legacy of cattle breeds. These were localised to the extent that many animals acquired vernacular names based on place association (Evans and Yarwood, 1995). Most were dual purpose animals, producing both milk and meat, and they became well-adapted to local conditions. From such diversity emerged dominance as Bakewell popularised one particular breed, the Longhorn, so that it became a common sight across the UK by 1800. This superiority was soon lost to the Shorthorn breed, itself derived from Durham and Teeswater cattle breeds (Walton, 1984), although it remained dual purpose. Post-war specialisation encouraged a divergence between dairy and meat functions, with productivist 'improvements' all based on the introduction of livestock from continental Europe into breeding programmes.

i) Dairy Cattle

In the dairy sector, one breed has come to dominate, namely the black and white Holstein-Friesian cow. This breed has a natural ability for exceptional lactation which has been fused with human energies in devising dietary regimes, medical intervention and automated milking procedures to produce a beast which is, contradictorily, of genetic purity but of cultural-natural hybridity (Whatmore, 2002). It is here that an assessment of sustainability can be brought sharply into focus. Pushing milk yields of Holstein-Friesians to the absolute maximum, anything up to 12,000 litres per annum, has had negative ramifications for the classic triumvirate of sustainability:

- Economic: the costs of dairy farming have risen sharply with investment in technology to support output from the breed (computerised milking parlours, bulk storage, insemination straws), yet production levels have far outstripped consumer demand (which has further fallen due to health worries about fat intake). In 1980, the EU was 222 per cent self-sufficient in milk (Bowler, 1986). Prices have tumbled and commentators are now talking of an exodus from dairying. In their recent study, Colman and Yaqin (2005: iii) conclude that "Not only did more farms cease production between April 2003 and April 2005 than had intended to do so in 2003, there was a higher concentration of more profitable and larger herds among those that quit."
- Social: farm businesses have shed labour to reduce costs, leading to the disappearance of specialist stockmen on many family labour-based farms. Those remaining on the farm are committed to at least a twice daily workload, exacerbated by increases in herd sizes. Indeed, studies have noted that the incidence of adoption of farm diversification amongst dairy farmers is lower than average because of their workloads (Evans *et al*., 2002b; University of Exeter, 2003). In one survey, a farmer remarked to us that milking three times a day, as opposed to two, put extra strain not only on the farm personnel but simultaneously, to use his words, "hammered the cows". This is without the use of the genetically engineered hormone recombinant Bovine Somatotropin (rBST), produced by Monsanto in the USA under the name 'Posilac', to boost lactation. It is banned within the EU, unlike the USA, because of concerns

about the physiological and psychological welfare of dairy cattle as well as human health effects. Holstein-Friesian cows further require a high protein diet to sustain high lactation. Rendered livestock remains provided a convenient nutritional source, yet this short-cut has been established as the cause of the UK outbreak of 'mad cow disease' (BSE).

- Environmental: the clustering of large amounts of animals in small spaces has raised problems with point-source pollution from slurry. More widespread has been the impact on landscapes and habitats of a switch from hay cropping to nutrient rich silage production. Holstein-Friesian cattle require silage feed to maintain a high milk yield. A sharp decline in grassland species diversity has been observed over a relatively short period of time.

If sustainability is not considered to be a three-legged milking stool, but rather a four-legged chair which includes political sustainability, then the persistence of milk quotas in the agrarian policy landscape is a constantly embarrassing reminder of the inability of government to regulate the output of milk derived primarily from Holstein-Friesians. With the dairy sector clearly failing to deliver sustainability to farmers, animals, consumers, politicians and nature, it is not difficult to see where the stumbling block lies.

ii) Beef Cattle
There is more breed diversity within the UK beef than dairy sector, although a group of breeds commonly known as 'continental cattle' have come to dominate production systems at the expense of indigenous British breeds. This group includes Charolais, Simmental, Belgian Blue and Blonde D'Aquitaine breeds which are of larger size and weight than most British breeds. They also tend to have a carcass conformation with less fat, as demanded by the modern consumer. For example, the British Charolais Cattle Society promotes its breed as 'the value-added breed within the beef sector' due to the qualities of carcasses from calves where a Charolais has been used as a 'terminal sire' (apparently, according to proponents of the breed, they 'weigh and pay'). The quid pro quo is that such cattle require high protein diets and supplements containing iodine, selenium and copper to perform to the level claimed for them. The need also for medical treatments, calving difficulties and the narrowness of genetic stock derived from a few beasts that have been declared breed 'champions' at points in the past are all issues for future sustainable development of the beef sector.

iii) Rare Breed Cattle
One consequence of such fundamental change is that there are livestock breed 'losers' at a disproportionately high rate for every 'winner' breed in the agro-industrial food production system. Thus, there are now more breeds of cattle within the most endangered categories on the RBST's survival watchlist than any other type of farm animal. Table 4.2 shows that there are no less than eight breeds of cattle whose numbers are so low that they are 'endangered' or in 'critical' danger of extinction. It is interesting to note that, in 1996, only five breeds of cattle had this status. The addition of three cattle breeds to these groups by 2006 is partly a

function of redefinition of rareness by RBST (the 'original populations'). However, it demonstrates that many breeds of cattle continue to be marginalised, even within what some commentators consider to be a multifunctional agricultural regime (Wilson, 2001) and despite the presence of an organisation whose prime mission is to save them. For example, numbers of Shetland and Vaynol cattle have not improved over the ten years monitored in Table 4.2, the Irish Moiled breed has improved slightly, but the numbers of breeding adult females for the two Shorthorn breeds indicated have fallen dramatically from in excess of 750 in 1996.

Table 4.2 indicates that RBST also identify cattle as 'rare' based on the identification of an endangered 'original population'. The purpose is to distinguish those cattle within a breed that look similar yet have a greater intrinsic genetic purity from others that bear the same name. This serves to demonstrate the primacy of genetics within the work of RBST. However, it should be noted that a second discourse is emerging to cross-cut the sustainability debate centred on factors of locality and tradition. Importance is becoming attached to 'Traditional Breeds' that may or may not be rare. For example, there is increasing interest in the keeping of Devon Cattle (Ruby Red) in the south-west of England (Yarwood, 2006). This breed is not genetically rare but these cows are, rightly or wrongly, becoming viewed as more sustainable due to their local geographical associations. Ironically, American Milking Devon cattle in the USA have more original genetic material than those

Table 4.2   The RBST's 'Rare Breed Watchlist 2006' and the situation in 1996

CATEGORY Adult females, Upper limits	BREED					
	2006			1996		
	CATTLE	SHEEP	PIGS	CATTLE	SHEEP	PIGS
CRITICAL cattle <150 sheep <300 pigs <100	Northern Dairy Shorthorn	Boreray		Irish Moiled	Castlemilk Moorit	British Lop
	Whitebred Shorthorn	North Ronaldsay		Shetland	Norfolk Horn	Tamworth
	Vaynol			Vaynol		Large Black
	Chillingham			White Park		Middle White
	Aberdeen Angus (OP)					
ENDANGERED cattle 250 sheep 500 pigs 200	Shetland	Castlemilk Moorit	British Lop	Gloucester		Berkshire
	Irish Moiled	Leicester Longwool	Tamworth			Saddleback
	Lincoln Red (OP)	Teeswater				Gloucestershire Old Spots

(OP = Original Population)

Source: The Ark, 1996 and 2006.

found in Devon, so that parochial-global tensions look set to emerge in the debate about sustaining livestock.

Sheep

There are greater numbers of sheep breeds than of cattle and, of course, there are more sheep in the UK in absolute terms. Nevertheless, the sector is again dominated by a few breeds. As a general rule, there is little current commercial interest in Britain with wool-producing breeds. The development of artificial fibres and international competition have meant that prices for fleeces continue to be depressed at 76 pence per kilo in 2005 (British Wool Marketing Board (BWMB), 2006a), falling further in late 2006 to around 65 pence per kilo (BWMB, 2006b). For illustrative purposes, the Cotswold sheep which has been bred for its exceptional high weight of fleece, and upon which the wealth of a whole region of England was founded in medieval times, will yield a fleece of between 5 and 6kg, although 2 to 3kg is a more typical average for many breeds. This price does not take into account operating costs of 27p/kg (BWMB, 2006a). The number of registered producers with the BWMB has fallen to 59,726, down from 74,995 in 2000, continuing a long-established trend (BWMB, 2006a). Product innovation and promotion using wool has done little to arrest this decline, with field trials using fleeces as a crop-mulch indicating the depths to which this part of the agricultural sector has sunk.

With a focus on meat production, the sheep sector has become reliant upon a few breeds, but with breeding fashions playing a greater role than with cattle. The UK effectively has a national flock stratified between 'pure-breeding' and 'finishing for market' functions. In recent years, the Scottish Blackface, Welsh Mountain and Swaledale sheep have come to dominate the upland breeding 'pedigree' flock due to their hardiness and fecundity. At the other end of the breeding spectrum, lambs for market are most commonly derived from a final cross with Suffolk sheep, but Texel (Netherlands), Charollais (France) and Rouge de l'Ouest (France) have become prominent since their importation into Britain from the 1970s. For example, according to their respective breed societies (but see also Pollott, 2005), Texels have 30 per cent and Charollais have 20 per cent of the terminal sire market primarily due to their ability to produce many lean lambs that quickly mature to market weight.

That some diversity of sheep breeds remains is due to at least four factors founded in policy, culture and science. First, the former headage basis of subsidy payments made to farmers under the CAP, with additional supplements for those in Less Favoured Areas, long encouraged high stocking densities (Fennell, 1979; Bowler, 1985b). Environmentally founded policy measures of the 1990s, such as extensification premia and the Moorland Scheme (which briefly but unsuccessfully offered farmers £25–£30 per ewe removed from the uplands and removed just 3900 sheep in its first year) did little to reduce stocking rates simply because they could not compete with the financial rewards that farmers could gain from headage payments (Winter *et al.*, 1998). The FMD outbreak did far more to reduce sheep numbers which fell from over 20 million head of breeding ewes in 2000 to 16 million in 2004 (Connor, 2005). The move to a Single Farm Payment under CAP in 2005, one that will eventually be based on farmed area rather than headage, should reduce the tendency towards overstocking and a steady decline in the national flock should follow (Connor, 2005). Second, there has been a cultural effect of breed loyalty by farmers in particular localities (Yarwood and Evans, 1999 and 2000; Yarwood, 2006). One example is the Rough Fell (or sometimes Kendal Rough) sheep:

> Although this breed is little known nationwide, it is enormously popular on its native mountain and moorland farms, embracing a large proportion of South Cumbria, parts of the West Riding of Yorkshire, North Lancashire and more recently, parts of Devon. This exceptionally hardy type of sheep has proved to be well fitted to endure the hardships of exposed and high upland mountains. (Rough Fell Sheep Breeders' Association, 2005, www.roughfellsheep.co.uk/)

A 'Rough Fell Sheep Country' map was published by its breed society in 2005 which can be viewed as maintaining and promoting the regional identity of the breed to both keepers and non-keepers alike. Third, the rise of a second pillar to the CAP has meant increased funding for rural development projects (Lowe *et al.*, 2002). Although not providing money for the keeping of sheep directly (dealt with under the first pillar), activities associated with the culture and social traditions of sheep have been supported. For example, an international Herdwick Sheep Shearing Competition in Cockermouth, Cumbria during July 2005 attracted 96 competitors, 3,000 spectators and 1,000 sheep! Fourth, some biotechnical developments in animal science have not been so easy to apply in sheep as in cattle so that a wider geographical spread of 'industrialised' breeds has been restricted. Artificial insemination of cattle has been one major way this industrialisation has been achieved in the dairy sector. However, it is far more difficult in sheep as the recto-vaginal method, where the operator inserts one hand into the rectum to manipulate the cervix and uses the other hand to insert vaginally the insemination straw, cannot be used. The SID, or 'shot in the dark', method using a pipette leads to a conception rate frequently less than 30 per cent (Pattinson, 2005).

Nevertheless, the history of marginalisation of primitive breeds such as the Soay, Boreray and North Ronaldsay sheep continues. These have already been displaced to the islands of Scotland due to their small carcasses, strong flavoured meat product and lack of flocking instinct in favour of the Scottish Blackface. In the case of the North Ronaldsay, it was removed from the pastures of the island and left to roam beyond the sea wall to eke out a precarious existence by grazing on seaweed (Evans and Yarwood, 2000). A new threat to primitive (and some other) breeds comes in the form of the National Scrapie Plan. Scrapie is a brain disease in sheep similar to BSE in cattle which the Government wants eventually (15 years is the minimum period identified) to eradicate over fears about its transmissibility across species boundaries and possible (though as yet unproven) impact upon public health. The plan attempts to ensure that only sheep naturally resistant to scrapie exist within the national flock. Genotyping methodologies have identified that sheep with certain genetic alleles, especially one known as VRQ after its combination of amino acids, are more susceptible to the agent thought to transmit scrapie. Unfortunately, hill sheep, and particularly primitive sheep with coloured coats and horns, most commonly possess the VRQ allele. Fears are expressed that this programme will reduce genetic diversity within the UK national flock whilst not eliminating scrapie as one strain of the disease has already been found to infect the most resistant ARR allele (Elliot, 2005).

Pigs

Breed diversity in the pig sector has long been a victim of agricultural industrialisation. Modern developments in pig production are summarised by Symes and Marsden (1985: 100):

> The livestock are thus reared and fattened in an enclosed, stable thermal environment; fan ventilation is necessary to prevent the build-up of toxic gases and reduce outbreaks of disease among the densely housed stock. With the aim of reducing labour costs, straw bedding may be dispensed with and slatted floors introduced to allow drainage of slurry into underground storage tanks. Farrowing sows may be 'crated' to prevent accidental crushing of their young and dry sows tethered or confined to cubicles to restrain aggressive social behaviour in high density conditions. In such conditions, the risk of epidemic disease within the herd is greatly increased and 'medication' of young stock with antibiotics is common. Intensification of the system is achieved not only through high density stocking but also by shortening the breeding cycle.

Such systems have heavily favoured two pig breeds; the Landrace and the Large White. Both produce lean, long carcasses which the modern market desires. Breeds producing fatty pork products have been disfavoured, some to the point of extinction. Hence, the most recent UK livestock breed extinctions include pigs such as the Cumberland, Ulster White, Yorkshire Blue, Oxford Sandy and Black (disputed) and Lincolnshire Curly Coat. Many of these were popular prewar, but this description of the Lincolnshire Curly Coat aids an understanding of why it became extinct:

> ... sturdy and hardy workers in its county in many cases have not altogether lost their appetite for well-cured and matured bacon fortified by a thick firm layer of fat such as the more southern townsmen dare not face (Layley and Malden, 1935: 115).

Unlike the other livestock sectors discussed, pig production has been left to the vagaries of free market economics, with a recurrent boom and bust trend, known as the 'pig cycle', discouraging the local breed loyalty amongst producers that is evident with other British livestock. Disease risks, discontinuation of feeding human food waste (pig-swill) and specialisation in other agricultural enterprises has led to the virtual elimination of small pig herds that were a characteristic part of the farmyard scene before the 1960s.

Policy Veneers

Although well-established as oxymoronic from a political economy perspective (Redclift, 1987), any mission to attempt the sustainable development of rural areas should not for practical, historical and ethical reasons ignore livestock. It is clear that adherence to the agro-industrial model can only ever support a limited number and type of livestock, but that this single trajectory of agricultural development has become less focused and more uneven, possibly since the 1980s (see Evans *et al.*, 2002a). A shift away from productivism towards a multifunctional agriculture suggests a re-emergence of livestock as critical components of a sustainable rurality. Roles in heritage marketing, speciality foods, farm-based tourism and environmental management have all been identified (Yarwood and Evans, 2000). Recent policy initiatives might be viewed as encouraging this process; an assertion which is worthy of further investigation taking the environmental management dimension as an illustrative example.

In the UK, the introduction of a suite of measures broadly known as agri-environmental policy has undoubtedly provided some incremental steps towards thinking about greater rural environmental sustainability, even if the actual longevity of their achievements can be questioned. In England, the diverse range of agri-environmental schemes that emerged from the mid-1980s and proliferated during the 1990s have now been largely consolidated within the Environmental Stewardship (ES) scheme. Until recently, one feature persisted in the design of agri-environmental measures – a neglect of livestock generally and breeds specifically. This was despite the experience of some other EU member states, which for some time have attempted to offer farmers support for specific breed-environment associations, such as the Rural Environment Protection Scheme (REPS) in the Republic of Ireland (Emerson and Gilmour, 1999; Yarwood and Evans, 2003). For example, in the REPS 3 version of this policy (superseded in 2007 by REPS 4), 'Supplementary Measure 3' was devoted entirely to the 'Conservation of Animal Genetic Resources (Rare Breeds)' in recognition of the fact that:

> Local animal breeds play a significant role in maintaining the rural environment and represent a significant element of the cultural heritage of farming in Ireland (Department of Agriculture and Food, Ireland, 2005: 37).

Breeders of Kerry, Dexter and Irish Moiled (Maol) Cattle and Galway Sheep were able to receive 200 Euros per livestock unit (equivalent to 1.0 for cattle and 0.15 for sheep) kept. In contrast, the UK's Entry Level Scheme (ELS) of ES favours reference to livestock where they should be managed to prevent erosion, pollution or woodland grazing. One exception is option EK5 which offers lowland farmers payment to re-establish 'mixed stocking' practises to encourage the breeding of farmland birds. Upon its launch, even the Higher Level Scheme (HLS) of ES, designed to deliver more conservation from farmers, was primarily concerned with the negative impacts of livestock through overgrazing, illustrated by its detailing of various management situations in which livestock should be excluded. A 'Shepherding supplement' (HL16) to allow certain distributions of livestock and a 'Supplement for difficult sites' (HR7) to compensate for the extra management costs where livestock safety might be compromised were the only features in the original package that could be remotely considered as concessions to livestock within ES. Not until July 2006 did UK agrarian policy make progress towards recognising the importance of specific livestock breeds within the ES flagship scheme. The Department for Environment, Food and Rural Affairs (Defra) announced that extra money was being made available to farmers through HLS to support the keeping of 'native breeds at risk'.

The aim is to deliver conservation objectives relating to habitats and, at long last, the 'genetic heritage' of the animals themselves.

Another cornerstone in the UK's drive towards achieving sustainable environmental management is Biodiversity Action Planning (or BAPs). Briefly, this initiative emerged from the 1992 Rio Earth Summit and the UK's signing of the Convention on Biological Diversity (CBD). A strategy was published, setting out clearly the steps that should be taken to improve the abundance of 382 (rare) species of flora and fauna (known as Species Action Plans or SAPs) and the incidence of 45 of their habitats (Habitat Action Plans or HAPs) (DoE, 1994). This was deemed to be a more proactive approach than the 'old-fashioned' one previously enshrined in the 1981 Wildlife and Countryside Act of simply listing and legally protecting species already on the brink of extinction (Gilg, 1996). BAPs comprise costed plans that involve conservation organisations working together in partnership to achieve results. Among the advantages of BAPs, Adams (2003) notes partnership working, job creation and fund-winning in a new enterprise culture. Yet, he questions their real achievements, as not one SAP target had been reached by 2001 despite their detail, and also their practical relevance to decision-making landowners. Consumption of resources has been significant and the preparation BAPs have certainly kept conservationists busy, to the extent that Adams (2003: 180) critically describes the process as "... a monster, sucking in resources ..."

With the aim of providing detailed information on species recovery, a key question arises: how does this biodiversity planning engage with the relationship between livestock (breeds) and the environmental conditions that their presence are likely to bring about? One headline species which grows on moderately grazed limestone grassland in northern England is *Cypripedium calceolus*, better known as the Lady's Slipper Orchid. Its SAP acknowledges that this large and colourful orchid is in decline for reasons of "habitat destruction due to increased grazing pressure", alongside picking by collectors. Table 4.3 summarises the actions to be taken as part of the recovery plan.

It is evident from Table 4.3 that the SAP for Lady's Slipper Orchid says nothing about livestock, despite overgrazing being identified as a key cause in its decline. The statement about 'appropriate methods of habitat management' quickly devolves down responsibility for 'on the ground' delivery to partner organisations, in this case, Natural England (for an example of this approach in practice, see English Nature, 2002). With only one plant remaining 'in the wild', it may be thought understandable that grazing regimes are not discussed in detail as so much restoration work needs to be done to re-establish this plant. However, other SAPs for more widespread species such as Irish Lady-stresses (*Spiranthes romanzoffiana*), where stocking densities, timing and a shift in grazing type from extensive cattle to intensive sheep are identified as instrumental in its decline, talk vaguely about 'establishing suitable grazing regimes' even though the future suggested research programmes reveal that little is known about what these might be! Adams' (2003) remarks about BAPs as paper exercises that merely confuse and distract conservationists are thrown sharply into focus.

Table 4.3   The Species Action Plan (SAP) for Lady's Slipper Orchid
(*Cypripedium calceolus*)

Action type	Summary of Action
Species management and protection	Ex-situ conservation
	Enforce plant protection
	Increase genetic diversity
	Maintain wild stock
	Restore populations to suitable sites
	Maintain national seed bank
Advisory	Ensure landowners are aware of appropriate methods of habitat management
Future research and monitoring	Survey former and potential sites
	Research seed storage methods
	Ensure information is passed to national databases
	Provide information to global databases

Source: Derived from Species Action Plans published by the Joint Nature Conservation Committee (JNCC).

Future Livestock

This discussion has attempted to illustrate that livestock have largely been neglected from thinking about rural sustainability. Where included, livestock are usually discussed in vague terms or cast as negative agents by virtue of overgrazing in the project to (re)establish diverse species assemblages. The main systems for the delivery of policy aimed at securing rural sustainability in the UK, especially in environmental terms, have barely started to embrace livestock breeds. This is despite the existence of some enabling frameworks at the EU level, such as Article 14 of Commission Regulation 455/2002 on endangered 'local breeds' (Yarwood and Evans, 2003). The decline in livestock diversity has been one major consequence of the post-war industrialisation of agriculture. Certain breeds may enjoy short-lived fashionableness under such conditions, now regarded as relatively unsustainable, but only a narrow range of breeds will be supported. The same is true where livestock have been selectively recommodified as objects of tourism or as quality food products during the drive by farmers to diversify their businesses (Yarwood and Evans, 2000). The existence of a sustainable stock of farm animals very much depends upon a clear definition of purpose, because the relatively small membership of the RBST demonstrates the limited power of intrinsic appeal to conserve livestock breeds.

Most potential for influence on the survival and abundance of livestock lies in the arena of environmental quality. The analysis presented here shows that most current environmental policy treats livestock as a broad brush problem to, rather than a cause of, biodiversity. It cannot be denied that overgrazing has been environmentally pernicious, although a more correct assertion would be to qualify this statement by locality and breed type. Some conservation organisations, such as the RSPB, have had to turn to the concept of 'flying flocks', as piloted under the Grazing

Animals Project (GAP) in the form of Local Grazing Schemes (Grayson, 2001), to (re)establish desirable grazing regimes on high nature value sites. This is indicative of the asymmetry between local environmental management and livestock (Evans *et al.*, 2003). Beyond redistribution, to embed livestock breeds into rural sustainability thus relies upon making explicit links between grazing habits of individual breeds and environmental outcomes. This important work has begun but remains nascent due to the sheer number of breed-management permutations that can exist and the limited numbers of both animals and humans that can take part in this research (Small *et al.*, 1999; Yarwood and Evans, 2003). Once such associations are understood, a firm foundation will have been established from which to enjoy the other cultural, aesthetic and ethical benefits that farm animals can bring to future generations of those experiencing countryside.

References

Adams, W. (2003) *Future Nature* (2nd Ed.). London: Earthscan.

Bowler, I. (1985a) Some consequences of the industrialisation of agriculture in the European Community. In: Healey, M. and Ilbery, B. (eds), *The Industrialisation of the Countryside*. GeoBooks: Norwich, pp. 75–97.

Bowler, I. (1985b) *Agriculture Under the Common Agricultural Policy*. Manchester: Manchester University Press.

Bowler, I. (1986) Intensification, concentration and specialisation in agriculture: the case of the European Community. *Geography*, 71: 14–21.

Bowman, J. and Aindow, C. (1973) *Genetic conservation and the less common breeds of British cattle, pigs and sheep*. Department of Agriculture and Horticulture, University of Reading, study No. 13.

British Wool Marketing Board (2006a) *The Year in Summary*. Accessed at http://www.britishwool.org.uk/summary.asp?pageid=29.

British Wool Marketing Board (2006b) *Price Indicator*. Accessed at http://www.britishwool.org.uk/priceindic.asp?pageid=45.

Clutton-Brock, J. (1981) *Domesticated Animals From Early Times*. London: British Museum and Heinemann.

Colman, D. and Yaqin, Z. (2005) *Changes in England and Wales Dairy Farming Since 2002/03: A Resurvey*. Manchester: Centre for Agriculture and Food Resources, University of Manchester.

Connor, J. (2005) Outlook for the UK sheep sector. Paper presented to the Meat and Livestock Commission Outlook Conference, London, 26 January.

Department of Agriculture and Food, Ireland (2005) *Farmers Handbook for REPS 3*. Co. Wexford, Republic of Ireland: DAFF.

DoE (1994) *Biodiversity: the UK Action Plan*. London: Department of Environment.

Elliot, G. (2005) Scrapie – another way. *The Ark* 33, Summer: 20.

Emerson, H. and Gilmour, D. (1999) The Rural Environment Protection Scheme of the Republic of Ireland. *Land Use Policy*, 16: 235–245.

English Nature (2002) *Traditional Breeds Incentive for Sites of Special Scientific Interest*. Taunton: English Nature Somerset and Gloucestershire Team.

Evans, N., Gaskell, P. and Winter, M. (2003) Re-assessing the role of agrarian policy and practice in local environmental management: the case of beef cattle. *Land Use Policy*, 20: 231–242.

Evans, N., Morris, C. and Winter, M. (2002a) Conceptualising agriculture: a critique of post-productivism as the new orthodoxy. *Progress in Human Geography*, 26: 313–332.

Evans, N., White, K., Lovelace, D. and Storey, D. (2002b) *A Farm Study for Shropshire*. Centre for Rural Research, University of Worcester, Worcester: Report to Shropshire County Council and partners.

Evans, N. and Yarwood, R. (1995) Livestock and Landscape. *Landscape Research*, 20: 141–146.

Evans, N. and Yarwood, R. (2000) The politicization of livestock: rare breeds and countryside conservation. *Sociologia Ruralis*, 40: 228–248.

Fennell, R. (1979) *The Common Agricultural Policy of the European Community*. London: Granada.

Gilg, A. (1996) *Countryside Planning* (2nd Ed.). London: Routledge.

Goodman, D. and DuPuis, M. (2002) Knowing food and growing food: beyond the production-consumption debate in the sociology of agriculture. *Sociologia Ruralis*, 42: 6–23.

Grayson, B. (2001) *Local Grazing Schemes: A best practice guide*. Grazing Animals Project: Local Grazing Schemes Initiative.

Haezewindt, P. (2003) Investing in each other and the community: the role of social capital. *Social Trends*, 33: 19–27.

Holloway, L. (2004) Showing and telling farming: agricultural shows and re-imagining British agriculture. *Journal of Rural Studies*, 20: 319–330.

Lampkin, N. (1990) *Organic Farming*. Ipswich: Farming Press.

Layley, G. and Malden, W. (1935) *The Evolution of the British Pig: Past, present and future*. London: J. Bale & Co.

Lowe, P., Buller, H. and Ward, N. (2002) Setting the next agenda? British and French approaches to the second pillar of the Common Agricultural Policy. *Journal of Rural Studies*, 18: 1–17.

Morris, C. and Evans, N. (2004) Agricultural turns, geographical turns: retrospect and prospect. *Journal of Rural Studies*, 20: 95–111.

Pattinson, S. (2005) Artificial insemination in sheep. *The Ark*, 33, Autumn: 32–35.

Philo, C. and Wilbert, C. (eds) (2000) *Animal Spaces, Beastly Places*. London: Routledge.

Policy Commission on the Future of Farming and Food (2002) *Farming and Food: A sustainable future* (the Curry Report). London: Cabinet Office.

Pollott, G. (2005) *The Breeding Structure of the English Sheep Industry 2004*. London: Report to Defra, Imperial College London.

Redclift, M. (1987) *Sustainable Development: Exploring the contradictions*. London: Methuen.

Ricketts-Hein, J., Ilbery, B. and Kneafsey, M. (2006) Distribution of local food activity in England and Wales: An index of food relocalization. *Regional Studies*, 40: 289–301.

Ritvo, H. (1987) *The Animal Estate: The English and other creatures of the Victorian age*. Cambridge, Mass: Harvard University Press.

Robinson, G.M. (2004) *Geographies of Agriculture: globalisation, restructuring and sustainability*. Harlow: Pearson.

Rough Fell Sheep Breeders' Association (2005) *Description and History of Breed*. Accessed at www.roughfellsheep.co.uk/.

Small, R., Poulter C., Jeffreys D. and Bacon, J. (1999) *Towards sustainable grazing for biodiversity*. English Nature Research Report, No. 316. Peterborough: English Nature.

Symes, D. and Marsden, T. (1985) Industrialisation of agriculture: intensive livestock farming in Humberside. In: Healey, M. and Ilbery, B. (eds), *The Industrialisation of the Countryside*. Norwich: GeoBooks, pp. 99–120.

University of Exeter (2003) *Farm Diversification Activities: Benchmarking study 2002*. Exeter: Report to Defra by the Centre for Rural Research, University of Exeter and Rural and Tourism Research Group, University of Plymouth.

Walton, J. (1984) The diffusion of the improved Shorthorn breed of cattle in Britain during the eighteenth and nineteenth centuries. *Transactions of the Institute of British Geographers*, New Series, 9: 22–36.

Whatmore, S. (1997) Dissecting the autonomous self: hybrid cartographies for a relational ethics. *Environment and Planning D: Society and Space*, 15: 37–53.

Whatmore, S. (2002) *Hybrid Geographies*. London: Sage.

Whitby, M. (ed.) (1994) *Incentives for Countryside Management*. Wallingford: CABI.

Wilson, G. (2001) From productivism to post-productivism ... and back again? Exploring the (un)changed natural and mental landscapes of European agriculture. *Transactions of the Institute of British Geographers*, New Series, 26: 77–102.

Winter, M., Gaskell, P. and Short, C. (1998) Upland landscapes in Britain and the 1992 CAP reforms. *Landscape Research*, 23: 273–288.

Yarwood, R. (2006) Devon livestock breeds: a geographical perspective. *Report and Transactions of the Devonshire Association for the Advancement of Science, Literature and the Arts*, 138: 93–130.

Yarwood, R. and Evans, N. (1999) The changing geography of rare livestock breeds in Britain. *Geography*, 84: 80–91.

Yarwood, R. and Evans, N. (2000). Taking stock of farm animals and rurality. In: Philo, C. and Wilbert, C. (eds), *Animal Spaces, Beastly Places*. London: Routledge, pp. 98–114.

Yarwood, R. and Evans, N. (2003) Livestock, locality and landscape: EU regulations and the new geography of Welsh farm animals. *Applied Geography*, 23: 137–157.

Yarwood, R. and Evans, N. (2006) A Lleyn sweep for local sheep. *Environment and Planning A*, 38: 1307–1326.

Chapter 5

A Study of the Motivations and Influences on Farmers' Decisions to Leave the Organic Farming Sector in the United Kingdom

Frances Harris, Guy M. Robinson and Isabel Griffiths

The expansion of organic farming in the United Kingdom (UK) from the early 1990s has reflected the growing concern amongst the general public with the quality of food. This has been fuelled by concerns about food safety, the welfare of farm animals, the sustainability of rural economies and the negative impacts of productivist agriculture on the environment (Lang *et al.*, 2001). So foods that can be shown to have some measure of being 'green', 'healthy' or 'environmentally friendly' have become more attractive to certain consumers. This attraction also reflects growing differentiation in the market place, whereby increasing numbers of consumers associate themselves with the purchase of 'green' products, including organic foods (Bell and Valentine, 1997). Moreover, both producers and retailers have viewed the development of 'green', 'quality' and organic foods as an opportunity to add value to their products.

Organic farming in the UK in the last two decades has frequently been portrayed as one of agriculture's most successful sectors, with significant expansion from a small base and opportunities for farmers to profit from the growing public concern for the quality and healthiness of the food they consume. This view is supported by various statistics. The organic market in the UK in 2004 was worth over £1 billion, an increase of 10 per cent on the previous year. The area of organically managed farmland in March 2004 (nearly 700,000 ha) was over twelve times that in 1997 and there were just over 4,000 organic producers of whom two-thirds were producers of livestock and livestock products. The UK is now the fifth largest producer of certified organic produce. Yet the much-vaunted success of organic farming hides a more complex dynamic in which there has been a growing turnover of organic farmers and a halt to the rate at which farmland is being converted to organic production.

This chapter examines the phenomenon of movements into and out of organic farming in the UK. It provides background on the recent slow-down in the expansion of organic farming in the UK, considers possible reasons for this development, and tests these reasons on the basis of recent data from two of the leading organic certification bodies. Its analysis is based partly on interviews and questionnaire surveys with farmers in southern England who have recently left organic farming.

Conversion to (and Reversion from) Organic Farming in the United Kingdom

To meet and stimulate the growing demand for organic produce, governments throughout the Developed World have introduced supportive measures, as exemplified by the introduction of subsidies for conversion to organic farming in the UK from the early 1990s. In the European Union (EU) conversion from conventional to organic farming is regulated through standards set in Regulation 2092/91. In complying with this Regulation, UK farmers wishing to farm organically register with one of ten certification agencies, which set standards, inspect farms, and (subject to satisfactory compliance with regulations) licence farmers to produce and sell organic produce. Farmers must comply with organic standards for two years prior to gaining full organic status. Land under conversion may suffer from lower productivity than other farmland (due to the inability to use oil-based fertilisers, pesticides etc.), and the produce does not command the price premium associated with organic status (O'Riordan and Cobb, 2001). Therefore, farmers on farms in conversion can suffer a difficult time financially. In order to offset this, the UK government offers financial support for those wishing to convert.

In 1999 the government doubled the funding available for conversion through the introduction of the Organic Farming Scheme (OFS), replacing the former Organic Aid Scheme (OAS) (Midmore *et al.*, 2001). The new scheme required a minimum contract period of five years of organic management during which time, like the OAS, producers had to be registered with an approved organic certification body. Initial funding aimed to encourage an additional 75,000 ha of new conversion (MAFF, 2000). Subsequently, £140 million was targeted at extending conversions between 2000 and 2006. This provided payments for conversion on the basis of £450 per ha per annum over five years for arable land, £350 per ha for other improved land, and £50 per ha for unimproved land. The uptake of the OFS was substantial, with the initial budget of £12m for two years being spent in the first six months of the scheme being opened. The amount of land entering conversion tripled from 60,000 ha in 1998 to 180,000 ha in 1999 (Soil Association, 2003).

This ongoing government support contributed to a rapid acceleration in the number of organic farmers and in the area devoted to organic production in the 1990s. Subsequently, though, there has been a fall in both the organically farmed area and the numbers of organic farmers (Figure 5.1). This masks a significant turnover amongst organic farmers in which there are both new entrants to the sector and producers withdrawing from organic certification (Ilbery *et al.*, 1999; Soil Association, 2003). It is this turnover that is investigated in this chapter.

Methodology

The analysis utilises data provided by two of the largest of the ten organic certification bodies operating in the UK, namely the Soil Association and Organic Farmers and Growers.[1] This was supplemented by in-depth interviews with a sample of farmers who have withdrawn from organic certification. Together, the two certification bodies certify 80 per cent of organically managed land in the UK. Data were

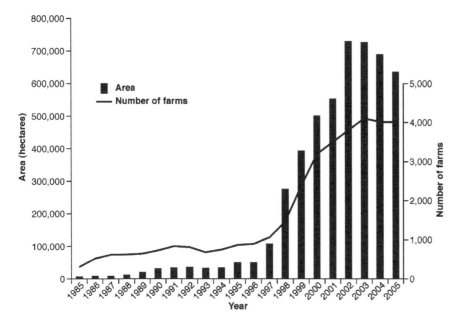

Figure 5.1 **Development of organic land area (hectares, left axis) and number of farms (right axis) in the UK**

gathered in a three-stage process, with each tier of investigation providing more finely grained detail:

- Both certification bodies provided information on the number of farmers who had left organic certification between January 2000 and December 2003. This information provided an overview of movements out of organic farming throughout the UK.
- One certifying body also provided further details about those leaving organic certification, and therefore farm characteristics as well as farmers' reasons for not renewing certification are available (for a group of 321 farmers). Our research aimed to focus on farmers who had voluntarily left the organic sector, rather than those who had been forced to do so through infringement of organic farming standards, or non-payment of licensing fees (of whom there were 68). Once these farmers were excluded from the larger sample, we were left with a sub-sample of 253 farmers who had voluntarily left organic farming, 176 of which provided information about their reasons for leaving the organic farming sector. This data has been analysed to identify general trends.
- The more generalised data from the certification bodies was supplemented by detailed interviews with 22 of the farmers who had left organic farming between 2000 and 2003. Farmers were contacted by letter and invited to participate in the research, based on the following criteria:
  - Certification had ceased between 2000 and 2003;

- Location and feasibility to interview. Focus was placed on livestock enterprises in southwest, western and southern England as this area has the main concentration of the country's organic farms (Figure 5.2a) and where the main focus of organic production is on livestock (Figure 5.2b);
- Farm size: primarily over 20 hectares, so that the focus was on commercial rather than hobby production;
- Certification was relinquished by the producer and not instigated by the certification body for infringement of regulations or non-payment of license fee.

Face-to-face interviews were conducted in early 2004. The interviews were carried out in a semi-structured manner, supplemented with unstructured dialogue that allowed the producers to cover issues of importance to them. The interviews were tape-recorded, which allowed further analysis after the interviews had taken place. The interviews also elicited basic information regarding farm management, involvement in the OFS, motivations and, especially, influences on the decisions to enter organic farming and then to relinquish organic certification.

The National Picture

Figure 5.3a shows that the largest concentration of the 321 farmers referred to above who had quit organic farming was in the South West and West Midlands, with these two regions accounting for 48 per cent of those leaving. Nearly 40 per cent of these farmers had holdings under 10 ha and 78 per cent farmed less than 50 ha. The area lost to organic farming from these terminations of certification were generally quite small: 65 per cent of farmers had less than 10 ha under organic production and 77 per cent had less than 10 ha in the process of organic conversion. As shown in Table 5.1, where a clear reason for leaving organic farming was provided to the certifying body, the most frequently cited single reason was because the farm business had ceased to be viable, with insufficient turnover and/or the farmer had been made bankrupt or the farm was sold. In some cases there were specific problems with the certification process. This included the farmers deeming the standards required as being too high and/or the fees being too high, as well as the certifying body taking action because of unpaid fees or through manifest infringements. Transfers to other certification bodies accounted for 18.2 per cent of those giving up certification. Hence this land usually stays under organic management. More minor reasons for quitting certification were the impacts of the foot-and-mouth disease outbreak of 2000/1 and technical or regulatory impacts. There were a few farmers who explicitly cited lack of premium prices for organic produce or lack of a market/customer base as the key factors behind their decision to quit, rather than just stating their farms were 'not viable'.

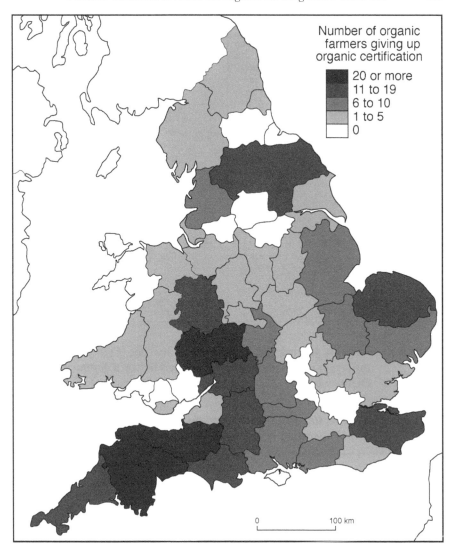

Figure 5.2   Number of farms leaving the two largest certifying bodies operating in the organic sector in England, from January 2000 – December 2003

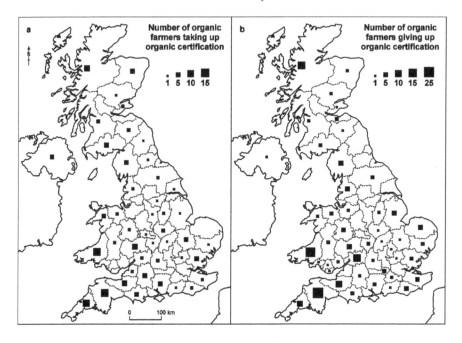

Figure 5.3   **Farmers joining (Figure 5.3a) and leaving (Figure 5.3b) one of the larger certifying bodies in England between January 2003 and October 2004**

Table 5.1   **Reasons for withdrawal from one of the major organic certifying bodies operating in England, 2000–2003**

Category	n	%	Stated reason	n	%
Stated reason	176	54.8	Not viable	48	27.3
No stated reason	77	24.0	Transferred to another agency	32	18.2
Terminated by soil Association	68	21.2	Fees too high	29	16.4
			Farm sold/tenancy relinquished	17	9.7
			Certifying body standards too high	16	9.0
			Personal reasons	10	5.7
			Foot-and-mouth disease	10	5.7
			Technical/registration issues	7	4.0
			Lack of market	7	4.0
	321	100.0		176	100.0

Reasons for Leaving Organic Production (Exit Themes)

The more focused information concerning the detailed reasons why farmers chose to leave organic certification is based on interviews with a sample of 22 organic farmers. Eighteen of the 22 producers remained in farming after they left organic certification. Of the four that did not, two producers took up other businesses, one retired and the remaining farmer rented out the organic land.

Four main sets of reasons emerged as underlying the decisions of these farmers to quit organic certification:

- Financial reasons: lack of market demand and low prices for organic produce;
- Negative experiences of the certification and inspection process and/or the certification fees;
- Negative experiences of the organic system on the farm;
- Other factors, including the foot-and-mouth crisis, personal circumstances and the distance to certified organic abbatoirs.

a) Financial Reasons

Financial motivations underpinned entry to organic production. For 17 of the 22 'quitters', entering organic farming had been regarded as a way to increase income. This was phrased in various ways, and included: to take advantage of high organic prices, to enter 'new' markets, as a financially advantageous alternative to intensification, to reduce the cost of inputs, to take advantage of growing demand and, as a subsidiary factor, to obtain conversion grants. In a separate survey of 40 continuing organic farmers surveyed in Devon and Cornwall in 2005, two-thirds of those who had joined under the OFS stated that they were attracted by the payments available under this scheme (Shorland, 2005). In general those leaving organic farming stated that their financial expectations had not been met. Hence 36 per cent of the quitters had not even completed their minimum contract period of five years for the OFS and had to pay back the subsidy they had been given (in one case amounting to £22,000). They reported problems related to lower prices and lower market demand than they had expected. Indeed, all the farmers who had reached the stage of selling organic produce complained about insufficient demand and low prices, especially for milk, where there have been sustained decreases in gross margins for organic milk from 1998 onwards (Lampkin and Measures, 2001).

It was the failure of the farm businesses to achieve a sustainable financial position through organic production that was the principal reason why the majority of interviewees had left organic certification. This confirms the findings made by Rigby *et al* (2001), who looked at reasons for a sample of 35 organic horticulturalists who left the sector between 1990 and 1998 (though the lack of demand/markets was crucial).

The key financial issues were being unable to find a market for their organic produce and being unable to find price premiums sufficient to make the organic system viable. Irrespective of how long they had been in organic farming, all the producers experienced the effects of the expansion of the sector post-1999, principally greater competition and growing pressure on profit margins. It was these reduced profit margins or the lack of expected profits that were the principal factor leading to over one-quarter of organic producers in a survey in Devon and Cornwall stating a desire to withdraw from organic certification (Shorland, 2005).

A significant proportion of those leaving the organic sector had anticipated that the transition to organic farming would be an easy one requiring little change. This perception, coupled with the high demand for organic food at the time they entered conversion, lead the producers to expect a market to be there at the end of the conversion period. This then lead to disappointment when the markets did not materialise or returns were lower than expected.

It can also be hypothesised that producers with financial problems are more likely to have looked at conversion to organic production as an opportunity to increase farm income and save an unviable farm business. These can be termed 'pragmatic' organic farmers (see Fairweather, 1999), and it might be argued that this pragmatism extends to a greater willingness to jettison organic farming when difficulties arise in this sector. If there are a large number of these 'pragmatic' farmers then there may be a greater exodus from the sector now as the producers who commenced conversion after 1999 came to the end of their minimum five-year contract with the OFS during 2004. Moreover, adverse prices may lead to more committed organic farmers also leaving. This prospect is supported by the results of Shorland's questionnaire survey of 40 organic producers in Devon and Cornwall in 2005, 35 of whom were producing organic livestock. Nearly two-thirds reported that profits had fallen or had remained unchanged since 1999 and over one-quarter claimed to be actively considering leaving organic certification.

b) Negative Experiences of Certification and Inspection

Nearly half those farmers sampled had problems with annual certification inspections. The inspectors were frequently regarded as being 'unsympathetic'. However, there was a great variation in experience with respect to inspections, for example positive experiences with inspectors were often critical in encouraging producers to remain in organic production. This repeats findings made with respect to the operation of agri-environment schemes (Morris and Potter, 1995).

Half of the quitters referred to the certification body in a negative fashion. This was partly linked to the general bureaucracy of organic production, and especially to the cost of certification fees, which seemed to rise whilst prices went down. There were additional problems linked in some cases to compliance with organic standards, where the charge was that there was 'too little consideration from inspectors'. The perceived ease of transition may also have meant that producers did not properly plan the conversion in terms of the physical and financial changes that would occur, nor perhaps did they adequately familiarise themselves with the organic standards of their certification body. Almost half of those interviewed referred to their discomfort and disagreement with the certification and inspection process. Hence a good relationship with the certifying body seems a crucial underpinning to 'staying power' in the sector.

Certification fees were mentioned as problematic by producers from a range of farm sizes, and not just the smaller units. Certification fees were felt to be too high in relation to output (fees of the two certification bodies involved had also increased). It must be stressed, though, that there is often a range of factors involved in the decision to quit, linked to a dominant issue such as finance.

c) Problems with the Organic System on the Farm

The majority interviewed seemed unprepared for problems posed by conversion, including some who referred to issues relating to animal welfare and a perceived greater difficulty of spreading risk under organic farming systems. Typical of the majority view amongst those farmers interviewed was: "I went into it without thinking too much. I thought I was an absolutely ideal person, but the more I got into it the more I realised I wasn't!" Indeed, a common perception amongst the interviewees was that little change would be involved in going organic, that the farm was almost organic already and/or that the organic system would mean less work. Hence they had viewed the transition to certified organic production as easy and requiring little change to the way they were farming. They had considered their farming system to be "almost organic anyway", "half-way there" or indeed "completely organic".

Several of those leaving organic farming had felt that their farms and ways of farming were ideally suited to organic production and that the transition to organic would involve minimal or no change at all. Furthermore, several of these producers were already in another environmental scheme such as Countryside Stewardship, and so organic farming seemed 'a logical progression' to them, or they stated that they were heading towards organic farming anyway. Hence there was a widespread under-estimation of the extent of changes involved in the transition to organic farming. In the interviews key issues that emerged as having been neglected or overlooked in the conversion to organic farming were: coping with new bureaucracy, regulations and inspection processes; technical aspects of organic farming; financial constraints during the conversion process; and marketing. Amongst those who had grown organic crops a minority referred to greater problems with weeds and the cost of employing additional labour.

Problems within the organic farming system included issues of animal welfare, with some of those who kept sheep commenting that they felt their flocks were healthier before they became organic. However, this was not a general feature, and some of those giving up organic certification commented on the good health of their livestock under the organic system. Some of the former small organic livestock producers felt that organic livestock farms needed to be larger to succeed: "you can't farm organically on 60 acres of grass monoculture", and "organic principles are correct but we can't implement a whole system". This reinforces the view that it has been smaller organic producers who have been most likely to withdraw from certification. The ability to diversify on the farm may be another factor, as 70 per cent of those producers remaining in organic farming in the Devon and Cornwall survey had some aspect of farm diversification (e.g. bed and breakfast accommodation, and off-farm employment for one or more members of the farm household).

d) Other Reasons

The interviews revealed that for most farmers leaving organic farming there were a variety of reasons underpinning the decision to leave organic certification. Whilst financial reasons were dominant, and were often compounded by negative views of certification or unrealistic expectations regarding what an organic farming system would entail on their farm, various personal circumstances and outside influences often affected the decision to quit. For example, although none of those interviewed had first-hand experience of the 2001 foot-and-mouth outbreak on their own farms, one-third referred to it as a factor in their decision to quit. They claimed that the disease in their local area had posed them problems of overstocking and lack of grazing. Organic certification may have seemed a less important issue in the context of just trying to stay in farming.

There were also miscellaneous problems referred to by some interviewees. These problems were only partly related to being in organic farming (notably that conversion to organic farming was not a 'saviour' of the farm business). Amongst those mentioned were family commitments and staff problems; personal indebtedness not directly related to organic farming; approaching retirement age; distance to organic abbatoirs; and being penalized for being in both Countryside Stewardship and the OFS. Other miscellaneous factors were primarily related to the three other broad reasons already discussed above.

Re-entry into Organic Certification

All of the producers in the survey were asked whether they might consider going back into the organic sector and how influential certain economic and social factors might be in influencing their decision. Nine of the 22 producers (41 per cent) interviewed said that they would never consider going back into organic production, no matter what the economic or other persuading factors might be. For those who might consider organic production again, the main influencing factor would be a rise in demand for, and prices of, organic produce.

An increase in the OFS payments did not appear to be highly influential, partly because many of the producers had received all of the entitlement under the OFS and would not be eligible for any more should they return to the sector. Moreover, whilst the OFS payments had been considered as offering welcome assistance, with one exception they were not regarded as the most critical factor when it came to making the decision to convert to organic status. However, for the smaller-scale producers, the payments had been a specific incentive.

There was some interest expressed in returning to organic farming if this involved a reduction in both paperwork and certification fees compared with previous experience, but these were only rated as highly important by two or three producers, which is perhaps surprising given that half the respondents were strongly influenced to come out of organic production by the 'high' certification fees.

Most (seventeen) of the producers in the survey had regarded the principles of organic farming as a factor in their decision to convert, and many still agreed with the ideas of organic farming despite having left certification. Five of the producers claimed they were not at all interested in the ideology of organic farming. There were seven producers who expressed very strong commitment to the ideology of organic farming.

Discussion

The organic farming sector in the UK has undergone a period of rapid expansion in the 1990s, followed by growing uncertainty regarding prices, markets and its future, with a minority of organic farmers giving up their organic certification. In choosing to study this minority, we wished to focus on reasons underlying decisions to quit, thereby gaining insights to the problems facing organic farming. Our initial research highlighted the fact that there has been a significant outflow of farmers from organic farming from 2000 onwards, especially amongst smaller farmers. Amongst the farmers leaving organic farming key elements in the decision to leave have been the changing economic fortunes of the sector, the farmers' unrealistic expectations when converting to organic farming and various negative experiences of farming organically, notably with the certification process, the lack of premium prices (especially for organic milk), difficulties in applying the principles of organic farming and some specific setbacks, e.g. related to the outbreak of foot-and-mouth-disease.

In addressing the end of the period of rapid expansion of organic farming, Smith and Marsden (2004: 346–7) argue that without further increases in consumer demand or significant increase in state subsidies then there may be a ceiling in organic production. Given this changing climate for organic farming, it is important to consider the challenges that those farmers leaving the sector have found insurmountable, and to assess whether they represent the beginning of a larger outflow or just a temporary 'blip'. Interviews with a small sample of farmers enabled us to go beyond the cited reasons on the standard forms compiled by certifying bodies and focus on detailed discussions of events leading to decisions to leave organic certification. In most instances, the decision to leave certification was the result of several pressures rather than a single reason. However, three general trends can be discerned from the interviews with those leaving the sector: economic reasons have been paramount; these have been compounded in some cases by difficulties associated with the certification process; and a large proportion of those leaving seem to have been ill-prepared for the requirements of farming organically.

The principal theme emerging from this survey is that of economics. Financial motivations to enter organic production were the most prevalent amongst the respondents, and indeed it was the failure of the farm businesses to achieve a sustainable financial position through organic production that was the principal reason why the majority of respondents left organic certification. The key financial issues were being unable to find a market for their organic produce and not being able to find price premiums sufficient to make the organic system viable.

In assessing reasons for farmers to become organic, Fairweather (1999) concluded that some farmers are motivated by 'the organic philosophy and ideology', which includes concerns about personal health, animal welfare and chemicals in food. He classified this group as 'committed organic farmers' (see also Burton *et al.*, 1997; 1998; 1999; Lampkin and Padel, 1994). However, he also identified 'pragmatic organic farmers', motivated by the price premiums on organic food. He claimed these would leave the organic sector if the price premiums went down. Rigby *et al.* (2000; 2001) claimed that, given the UK government's conversion grants and the premium prices for organic food that were available in the 1990s, it is likely that more 'pragmatic' than 'committed' farmers were converting to organic production. They hypothesised that once premium prices disappear, it is more likely that the farmers who converted for economic reasons will leave organic certification and there will be a 'down-sizing' of the organic sector. Their hypothesis and prediction are supported by the research herewith. The organic sector may indeed be looking at a down-sizing in the next few years if many of the 'pragmatic' organic farmers cannot make a sustainable living out of organic production, although even 'committed' organic farmers may be caught up in the price squeezes of the expanding organic sector, and be forced to leave organic farming.

Our study has focussed on a small group of farmers who chose to leave organic production, often for economic reasons. While their comments should inform the organic movement of potential problems, they are not necessarily the harbingers of a large-scale exodus from organic production. The organic sector may simply need a period of consolidation before further expansion. Most of the producers in this survey had an empathy with the aims and ideas of organic farming and considered that the way they farmed before they went into certification was almost organic. The new environmental focus of payments within the Common Agricultural Policy may provide a halfway point for these farmers. However, among these environmentally conscious farmers, there may be a great number of producers who would consider entering organic production in the future once the market has stabilised. This study indicates that their success will not only depend on a healthy supply and demand ratio, but also on a certification system that provides support and effective communication, whilst the producers have a realistic expectation of what organic farming will entail.

Note

1 The two largest organic certification bodies in the UK are Soil Association Certification Ltd and Organic Farmers and Growers Ltd. Together these account for 80 per cent of organic producers. The other certification bodies are Scottish Organic Producers Association, Organic Food Federation, Bio-Dynamic Agriculture Association, Irish Organic Farmers and Growers Association, Organic Trust Ltd., CMi Certification, Quality Welsh Food Certification Ltd., and Ascisco Ltd.

References

Bell, D. and Valentine, G. (1997) *Consuming geographies: we are what we eat.* London: Routledge.

Burton, M.P., Rigby, D. and Young, T. (1997) Modelling the adoption process for sustainable horticultural techniques in the UK. *Discussion Papers, School of Economic Studies, University of Manchester*, No. 9724.

Burton, M,. Rigby, D., Young, T. and de Souza Filho, H.M. (1998) The adoption of sustainable agricultural technologies in Parana, Brazil. *Revista de Economia e Sociologia Rural, Brazil*, 36: 199–222.

Burton, M., Rigby, D. and Young, T. (1999) Modelling the adoption of organic horticultural techniques in the UK. *Journal of Agricultural Economics*, 50: 47–63.

Fairweather, J.R. (1999) Understanding how farmers choose between organic and conventional production: results from New Zealand and policy implications. *Agriculture and Human Values*, 16: 51–63.

Ilbery, B.W., Holloway, L. and Arber, R. (1999). The geography of organic farming in England and Wales in the 1990s. *Tijdschrift voor Economische en Social Geografie*, 90: 285–295.

Lampkin, N.H. and Measures, M. (2001) *Organic farm management handbook*. Aberystwyth: Welsh Institute of Rural Studies, University of Wales, Aberystwyth.

Lampkin, N.H. and Padel, S. (1994). *The economics of organic farming*. Wallingford: CABI.

Lang, T., Barling, D. and Caraher, M. (2001). Food, social policy and the environment: towards a new model. *Social Policy and Administration*, 35: 538–58.

Midmore, P., Padel, S., McCalman, H., Isherwood, J., Fowler, S. and Lampkin, N. (2001) *Attitudes towards conversion to organic production systems: a study of farmers in England*. Aberystwyth: Institute of Rural Studies, University of Wales, Aberystwyth.

Ministry of Agriculture, Fisheries and Food (MAFF) (2000). *England rural development plan 2000–2006*. London: MAFF.

Morris, C. and Potter, C. (1995) Recruiting the conservationists: Farmers' adoption of agri-environmental schemes in the UK. *Journal of Rural Studies* 11: 51–63,

O'Riordan, T. and Cobb, R. (2001) Assessing the consequences of converting to organic agriculture. *Journal of Agricultural Economics*, 52: 22–35.

Rigby, D., Young, T. and Burton, M. (2000) Why do farmers opt in or opt out of organic production? A review of the evidence. Unpublished paper presented at the Annual Conference of the Agricultural Economics Society, Manchester 14–17 April, 2000.

Rigby, D., Young, T. and Burton, M. (2001) The development of and prospects for organic farming in the UK. *Food Policy*, 26: 599–613.

Shorland, J. (2005). The effects of the Organic Farming Scheme on organic farming in Devon and Cornwall. Unpublished Geography BSc Honours dissertation, Kingston University London.

Smith, E. and Marsden, T. (2004). Exploring the 'limits to growth' in UK organics: beyond the statistical image. *Journal of Rural Studies*, 20: 345–57.

Soil Association (2003) *Organic Food and Farming Report 2003*. Bristol: Soil Association.

Chapter 6

GM Farming and Sustainability

Bruce D. Pearce

Introduction

The use of genetic modification of our food has been around for centuries, since the first crossings and selections by our own predecessors of the wild ancestors of our agricultural animals and crops. With the developments in molecular sciences during the 1980s, our ability to genetically modify plants and animals took a giant leap forward. It was now possible to move genetic material between organisms that were not sexually compatible and even between kingdoms, i.e. between animal, plant and bacteria. This opened up a wide vista of opportunities for new and improved animal breeds and crop varieties. However, this technology, its utilisation and commercial-isation (over the past decade), although accepted in some parts of the world, has caused great concern and a polarisation of views within Europe and particularly the UK.

What is a GMO?

The Department for Environment, Food and Rural Affairs (Defra) has a definition for a genetically modified organism (GMO) on its website. It states *"A GMO is defined in the legislation as an organism, with the exception of human beings, in which the genetic material has been altered in a way that does not occur naturally by mating and/or natural recombination"* (DEFRA, 2004a). The website also sets out the government's policy by saying that *"The Government has concluded that there is no scientific case for a blanket ban on the cultivation of GM crops in the UK, but that proposed uses need to be assessed for safety on a case-by-case basis."*

The case-by-case basis of assessment is contentious as there are arguments as to whether the GM technology is safe. This is a much larger argument that will not be addressed in this chapter. For further information on both sides of this debate see the GM Science Review Panel (2003) and (2004) and I-SIS (1999). For this chapter I will ignore this issue and deal with GM crops on the case-by-case basis, although below you will see that the number of cases is very limited.

Where are we with GM Crops and Food in the UK Today?

In 2005 there were no Genetically Modified (GM) crops grown in the UK (DEFRA, 2005a); foodstuffs have to be labelled when the level of GM is greater than 0.9 per cent; supermarkets require their own label products to be GM-free and the public is generally against eating GM food. But was it always like this? The first GM food introduced to the UK was a tomato puree introduced in 1996 through Sainsburys and Safeway. The product was labelled as GM and was seen to have a benefit to the consumer as the decreased processing costs allowed a cheaper product than conventional puree. It was a great success and by 1999 was the dominant product on the market. Yet its demise was rapid and within only a few years, by the turn of the century, the product was no longer available. There are contesting views on the reason from this (for a history of the product see Soil Association, 2003 and Strategy Unit, 2003), but what was clear was that by 2000 the public had no appetite for GM food and the supermarkets were clearing them from their shelves.

What are the Claims for and against GM Crops?

The claims that were initially made from the pro-GM side of the debate were impressive. They were such claims as the greening of the desert, cereals that would fix their own nitrogen (and so need no fertiliser), salt-tolerant crops through to crops with increased yields and nutritional content, pest and disease resistance and tolerance to herbicides. Few of these claims have been brought to market and a decade later they have been toned down.

So what are the arguments in favour of and against the use of GM crops and its role in sustainable agriculture? With a world population estimated to be currently nearly 6.5 billion and to rise to be over 7.5 billion by 2020 (US Census Bureau, 2005) there is, and will continue to be an increasing challenge to feed this burgeoning population. The pro-GM side of the debate claim that the use of GM crops can be used not only to feed this expanding population but also to reduce the environmental impact of agriculture, to improve the health and welfare of this growing population and to increase the wealth of farmers that grow GM crops. From the start the GM-sceptics raised concerns about the commercialisation of products of the agricultural GM industry. They questioned the ability of GM crops to deliver the environmental benefits, the environmental safety of GM crops as well as the safety of the transgenic technology itself. There were also concerns about the power and motives of the multinational companies driving the development of GM crops. This debate began in the mid-1990s when the first commercial plantings occurred and continues today.

The World Scene for GM Agriculture

The global market value of GM crops in 2004 was US$4.7 billion and the area of GM crops grown was estimated to be 81 million ha. The United States accounted for nearly 60 *per cent* of this land area with 47.6 million ha, followed by Argentina (16.2 million ha), Canada (5.4 million ha), Brazil (5.0 million ha) and China (3.7 million ha). Romania, Spain (0.1 million ha each) and Germany (<0.05million ha) were the only European countries that grew GM crops in 2004. The principal crops that were grown internationally were soybean (48.4 million ha), maize (19.3 million ha), cotton (12 million ha) and canola/oilseed rape (4.3 million ha). The main trait that accounts for about two-thirds of the area grown was herbicide tolerance. The remaining third was pest resistance in the form of a *Bacillus thuringiensis* (Bt) gene (James, 2005).

There is a reason why these GM crops and varieties were brought to the market first. Not only was the technology there to produce the transgenic plants and varieties but also in the case of the herbicide tolerant crops the seed and the herbicide are sold as a single product. When seed of a Genetically Modified Herbicide Tolerant (GMHT) variety is bought an agreement is joined into that the purchaser will only use the herbicide sold to them by the company producing the seed. In the cases of glyphosate and glufosinate-ammonium, the herbicides that most GMHT crops are tolerant to, these herbicides were coming out of patent and so cheaper, equally effective versions of the product would be available. By tying the farmer into using only that chemical provided by the seed producer the profits for the agrochemical company could be maintained and the freedom of the farmer to source cheaper supplies curtailed.

Environmental Sustainability

So what are the impacts of growing these crops on the sustainability of agriculture? Let's take these two main traits and look at what the inclusion of them into crop plants claims to do. Basically both traits were aimed at improving the sustainability of growing these crops. It is claimed that agrochemical use will be reduced, with a resulting environmental benefit as well as reduced costs (economic) and improved social benefits through less exposure of the farmer to herbicides and pesticides.

However is this true? The US has the most long-standing and well-developed GM agricultural systems and grows GM crops that could be grown in both the UK (maize and canola/oilseed rape) and Europe (soybean). A comprehensive report by Benbrook (2003) using the USDA – National Agricultural Statistics Service agricultural chemical usage data investigated pesticide use over the first eight years of commercialisation of GM crops in the US (1996–2003). Benbrook shows that economically the introduction of herbicide tolerant soybean has been of benefit to the farmer with a 50 per cent reduction in the costs of herbicide per acre over the time of his report. Nevertheless, he goes on to show that in the US the introduction of GM crops has seen a modest increase in the volume of pesticides being applied to maize, soybean and cotton (both GM and conventional varieties). Between 1996 and 2003 there was an increased amount of herbicide use (70 million pounds) that was offset to a small degree by a reduction in pesticide use (51 million pounds). His analysis goes on and breaks the time frame between 1996 and 2003 into the early years (1996–98) and the later years (2001–2003), which give a different picture. In the early years there was an overall reduction of over 25 million pounds in the amount of pesticide used. These quantities did not continue in that direction for long and in the later years there was an increase of over 73 million pounds of pesticides applied. This is not surprising as the use of a single herbicide to control weeds over a period of time is likely to change the weed populations and result in the need for further or additional herbicides or applications. There is evidence that weeds are becoming resistant to the herbicide (Plant Managers Network, 2002) and that volunteer plants left behind from the GM crops are becoming a problem (Anon, 2004).

Benbrook shows there have been some successes with Genetically Modified *Bacillus thuringiensis* (GMBt) crops in a reduction in pesticide use. There is evidence from around the world that the levels of Bt being expressed by the GM crops can vary greatly and when at the lower level can lead to crop failures (Ho, 2005). There has also been increasing anecdotal evidence of resistance developing to Bt crops (GM Watch, 2004) that would require increased levels of pesticides to be reintroduced to control these pests.

We can see changes in pesticide use, but what are the environmental impacts? There has been much concern about GMBt and whether this trait will have adverse impacts on non-target organisms. The most famous and well publicised of these studies was on Monarch Butterfly (Losey *et al.*, 1999) which showed that the larvae living in weeds near maize fields could potentially be affected adversely by Bt maize pollen drifting on to the foliage of plant species favoured by the butterfly. These results have been questioned because they came from small-scale laboratory trials with high levels of Bt toxin expressed and where the larvae had no choice but to eat the foliage. Following this work, a number of other studies (Oberhauser *et al.*, 2001; Pleasants *et al.*, 2001; Sears *et al.*, 2001; Stanley-Horn *et al.*, 2001; Zangerl *et al.*, 2001) when taken together, suggest that the risks to monarch butterflies posed by current GMBt maize are not likely to be significant. This is a single example and there are others that show similar diverse findings with some nil and others negative effects.

When the environmental effects of growing herbicide tolerant crop varieties are looked at there is a great deal more robust and detailed information. Between 2000 and 2003 the UK government funded the largest ever field trials of GM crops in the world. These were known as the Farm Scale Evaluations (FSE) (DEFRA, 2005b) and were undertaken by a consortium of some of the UK's top agricultural and environmental scientists. The trials investigated the impact of GMHT winter and spring oilseed rape, sugar beet and maize on the environment. The general conclusions from the study were that the growing of GMHT crops had a deleterious impact on biodiversity. Specifically oilseed rape and beet showed a decline in the abundance of biodiversity within the GM crops while there was an increase in the GM maize when compared to conventional (Firbank *et al.*, 2003; Burke, 2005). The effect that gave the most concern was the reduction in weeds and weed seeds within the GM crops (weeds and weed seeds are important food sources and refuges for much biodiversity) and a conclusion of the work was that growing GM crops could have implications for wider farm biodiversity. Subsequent work monitoring the numbers of weed seeds in the soil within the FSE trial sites has further reinforced these concerns by showing that, two years after the GMHT crops had been grown, weed seed banks were significantly affected, with the potential for causing longer-term effects on other organisms (Firbank *et al.*, 2005).

An issue that came out of the FSE work was that it was the herbicide regime that could be used with the GMHT crops that had an effect on biodiversity and not the crops being GM *per se*. Another finding that came out of the FSE studies was that the abundance of biodiversity between the different conventional crops (oilseed rape, sugar beet and maize) was as great as that observed between the GM crops and their conventional comparison (Firbank *et al.*, 2003). This illustrates the impact that any agriculture has on the environment.

These results, although for a small range of GM crops (primarily herbicide tolerant), show that moving from one intensive regime of production, conventional, to an even more intensive system, GM, has a negative effect on our environment and is highly unlikely to be sustainable. Other studies and reviews (Fuller *et al.*, 2005, Hole *et al.*, 2005) on alternative methods of production, such as organic, show that these less intensive system of production can produce real positive benefits to the environment over and beyond the baseline conventional production system used in the FSE.

Economic Sustainability

As for environmental sustainability the picture is not clear and often contradictory. In a recent briefing the International Service for the Acquisition of Agri-Biotech Applications (ISAAA), an independent pro-GM organisation, claims that GM crops are delivering benefits to consumers and farmers alike and quotes a global potential gain by 2015 of $210 billion (James, 2005). P G Economics, on behalf of the UK Cabinet Office, undertook work more relevant to the UK. This was a theoretical study that concluded that the adoption of GM technology and the growing of GM crops in the UK could result in increased profitability for UK farmers (P G Economics Ltd, 2003). For wheat it suggested that a GMHT could produce savings of between £23–£36/ha and GM fusarium resistance wheat between £10–£15/ha. For sugar beet GMHT would save between £36–£109/ha (depending on source data). There would also be a yield increase resulting in an increase in gross margin of between £75–£150/ha. Figures for other crops are not as clear but the general thrust was that either through reduced variable costs or improved yields there would be financial improvements over conventional costs.

A report by the Economics Research Service/USDA (Fernandez-Cornejo and McBride, 2002) gives a more mixed analysis. While GMHT maize improved net returns GMBt maize had a negative impact. GMHT soybean had no significant impact on net returns while GMBt cotton had a positive impact. The data used for these calculations were obtained in the years 1997 and 1998. The work by Benbrook (2003) showed that these were the years of lower herbicide use and so the figures for GMHT crops are likely to be no longer as optimistic.

The Institute of Science in Society (I-SIS), an independent GM-sceptic organisation, holds a review on their website that contradicts much of the above information (Ching and Matthews, 2001). This says that soybean yielded significantly lower than conventional varieties, as did sugar beet and oilseed rape. Studies also showed that there was no financial benefit to producing GM soybean or maize. The study by Benbrook (2001) gives examples of maize between 1996 and 2001. He says that the overall financial impact on farmers of growing GMBt maize during this period was a negative one, although the figures varied from year to year (1996, 1997 and 2001 were profit years with 1998, 1999, and 2000 being losses). This was due to spending $659 million on seed in these years and only making an additional

$567 million from the crop. This resulted in a net loss of $92 million or a loss of $1.31 per acre.

Again financial sustainability needs to be taken on a case-by-case basis. The natural fluctuations in agricultural systems and world commodity markets also add to the variation in economic sustainability. However, the evidence shows that there are financial risks in growing any crop and a GM crop is no guarantee of reducing these risks; in some cases it can make things worse. The current reluctance of European and many other consumers to buy GM products also makes the finances of the producing GM crops questionable. Even the Cabinet Office Strategy Unit (2003) reported "*In the short term, negative consumer attitudes can be expected to limit the demand for products containing GM foods.*" It continues "*any net cost and/or convenience savings associated with the current generation of GM crops would be likely to be outweighed by the lack of a market, limiting their economic value.*"

Social Sustainability

After the initial success of the GM tomato paste (see above) the appetite of UK consumers for GM food and products rapidly fell away and GM products disappeared from UK supermarket shelves. In July 1999 Sainsburys Supermarkets were the first UK supermarket to remove GM constituents from their own brand products (Sainsburys, 2005). This was rapidly followed by most of the other leading retailers. Tesco (the UK's largest supermarket) states on its website that it does not have any own-brand GM-food on its shelves (Tesco, 2005) because their customers fail to see the benefits.

It was into this climate that the first GM crops were working their way through the approval process for planting in the UK. Then in November 1999, the UK government working with the farming and biotechnology industry body, SCIMAC (the Supply Chain Initiative on Modified Agricultural Crops), developed the concept and launched the Farm Scale Evaluations of GM crops in the UK. As part of the FSE agreement it was agreed that there would be no widespread planting of GM crops grown in the UK until the FSEs were completed (DEFRA, 2002). This in effect was a voluntary ban on the planting of GM crops in the UK until the 2004/05 growing season.

In 2000 the government established the independent Agriculture and Environment Biotechnology Commission (AEBC) to provide advice. The brief of the AEBC was to look at current and future developments in biotechnology which had implications for agriculture and the environment, and to advise the government on their ethical and social implications and their public acceptability. The AEBC's first report was on the FSEs (AEBC, 2001) and amongst its conclusions was for the government to commission an independent review of all information that will complement the results from the FSEs, to commit to an open and inclusive process of decision-making and improve understanding of the basis of public views. These three recommendations were transformed into what became known as the GM dialogue (Defra, 2004b). There were three linked strands to the GM dialogue: (i) The public debate (ii) A review of the scientific issues (the GM Science Review Panel) and (iii) A study into the overall costs and benefits of GM crops (Cabinet Office, Strategy Unit Report).

The three were to run concurrently and to produce reports to government. The latter two have been mentioned and information drawn on elsewhere in this chapter.

The Public Debate was a truly new phenomenon where a government was asking its citizens what they really thought about a critical issue of the day. It was called 'GM Nation?' and was launched in June 2003. A steering committee was established comprising members with a range of views on GM crops. Documentation and a website (http://www.gmnation.org.uk/index.html) were established to encourage the public to get involved; to participate and to organise their own debates. The debates ran between the middle of June and the middle of July 2003 and their findings were collated, analysed and written up by the steering committee and published in September 2003 (DTI, 2003). The report details that there were 675 meetings and that over 8000 people attended some form of event. The website received over 2.9 million hits and the debate steering committee received over 1200 letters or e-mails, with most from individuals rather than organisations. The debate report produced seven conclusions:

1. *People are generally uneasy about GM.* This unease related to both food safety and the environment but also covered a range of broader social and political issues (power of multinationals etc).
2. *The more people engage with GM issues, the harder their attitudes and more intense their concerns.* The more people got to know about GM the more their unease developed and the uncertainties that this produced hardened their views. Although they were willing to accept that there could be benefits to medicine and to Developing Countries.
3. *There is little support for early commercialisation.* There was a range of views on this, but over half the participants never wanted to see GM crops grown in the UK and many of the others wanted to see varying lengths of delay (although a more informed sub-sample did not share this view but still desired much more information and regulation).
4. *There is a widespread mistrust of government and multi-national companies.* A range of political issues was raised that resulted in the suspicion of the motives, intentions and behaviour of those who have control of the commercialisation of GM crops. There was great suspicion that the government had already made up its mind on GM crops and the debate was a smokescreen and sop to the public. Also, the perceived power of the multinationals created unease. There was a view that the companies were motivated overwhelmingly by profit rather than meeting society's needs. Even where there was acceptance of benefits of GM technology there was a doubt that companies would actually deliver them. There was also a suspicion of information from or science funded by these companies.
5. *There is a broad desire to know more and for further research to be done.* There was a very strong wish to be better informed about GM from sources that they felt they could trust. There was a desire for a body of agreed facts that could be agreed on by all sides of the debate.
6. *Developing Countries had special interests.* Initially participants assumed that GM technology might help developing countries to produce more food

and offer other benefits. As the debate developed, views diverged, with some feeling there were real benefits and others believing that alternative technologies would deliver better results.

7. *The debate was welcomed and valued.* Even though there was scepticism of government motives for the debate, the view of those who attended and participated was that they appreciated not only being able to give their own views but to hear the views of others.

So did this groundbreaking activity of public participation change anything? A Mori poll published in July 2004 (Mori, 2004), a year after the end of the debate, showed that 36 per cent of those questioned were against, with 13 per cent in favour of GM food and 39 per cent neither for nor against. What was startling was that 85 per cent thought we did not know enough about the potential long-term effects of GM food on our health, but over half thought that GM could have future benefits for the environment, consumers and the Developing World. So there is still not an overwhelming wish to buy or consume GM products in the UK, and without this consumer demand there is no market.

Conclusions

We are now ten years into the era of GM farming, but it is not possible to answer, for now or always, the question 'Is GM food and farming sustainable?' It is also difficult to answer whether current GM food and farming is sustainable.

The question in itself is false as it assumes all GM is the same, and as we have seen, different GM crops have different requirements and impacts. Although there are increasing areas of land being put down to GM crops throughout the world, the vast majority is still within the US, so we can use this as our model. The pesticide usage in the US is on its way up again after an initial drop with the introduction of GM in the mid-1990s. This has been attributed to the changes in weed populations within GM crops. The financial returns for GM crops have also been shown to be questionable, with increased profits in some years and reduced in others. On the whole, there was seen to be a reduction in profits. The UK FSE results and follow-on studies have shown that the negative impact of growing just one season's GMHT crops is considerable. Put all this against the total rejection of GM food in most of Europe. There is a limited market for these crops, and this will hold down their prices on the world markets. This suggests that this form of agriculture is not sustainable.

Within the UK context it is an easier question to answer. There are no GM crops grown currently in the UK and the likelihood of a market for them is remote. The public does not trust or want GM products, as borne out by the public debate and the lack of GM products on our supermarket shelves. They do not see any benefits in the current generation of GM crops and are well aware of the risks that are involved. The FSE results and the subsequent follow-on studies of biodiversity impacts suggest that GMHT crops will have a hard time gaining legal approval to be grown in the UK and EU due to their environmental impact and there is a limited need or market for a GMBt crop in the UK.

Second and third generations of GM crops may be more palatable to the consumer with increased nutrient content or abilities to grow in more challenging environments. However, these crops will still have to overcome the scepticism and mistrust of the public who have been through a range of food scares.

It is important to put GM within a whole farming system and to ask if it is the answer. GM is only a technology and it is unlikely that the use of a single technology will overcome the problems of creating a sustainable agricultural system. GM technology is also a single strand of a much larger area of science known as molecular biology (including such activities as the human genome project and marker-assisted breeding). This is an extremely interesting and powerful area of science that can increase our understanding and ability to deal with the problems of sustainability and by stopping short of GM can allay the fears of many of the GM-sceptics.

When looked at the system level GM crops over simplify the problem as they are generally dealing with a single issue (a pest, weed etc). For GM crops to be successful, their impact on the whole farm system has to be understood and addressed. The failure of the current generation of GM crops to do this has to make them unsustainable. The FSE results are a good example of this, as it is unlikely that the crops being GM *per se* had an impact on the biodiversity. It was the way in which the GM trait allowed the crop to be managed. The GM crops were very successful in what they were created to do, i.e. allow more effective weed control. The outcome of this was that weed control was so effective that it removed food sources for other organisms, with detrimental impacts on other parts of the system.

This reductionist approach to dealing with problems within systems is of concern. We are already aware that current agricultural practices have seen rapid decline in biodiversity on our farms in the past thirty years. A more intensive approach to agricultural problems is likely to increase these problems. There are alternative whole-system approaches currently being deployed that need to be further investigated and introduced. Organic farming is a whole-system approach to farming and is an increasingly popular method of farming and of food consumption. It has also already been shown to deliver many of the desired environmental benefits over current conventional agriculture (which the FSE results show are better than GM).

References

Agriculture and Environment Biotechnology Commission (AEBC) (2001) *Crops on Trial. A Report by the AEBC.* http://www.aebc.gov.uk/aebc/pdf/crops.pdf.

Anon (2004) Welcome to the world of unintended consequences, *Farmers Weekly*, August 27–September 2.

Benbrook, C.M. (2001) *When does it pay to plant Bt corn? Farm-level economic impacts of Bt corn, 1996–2001.* Idaho: Benbrook Consulting Services. http://www.biotech-info.net/Bt_corn_FF_final.pdf.

Benbrook, C.M. (2003) Impacts of genetically engineered crops on pesticide use in the United States: the first eight years, *BioTech InfoNet, Technical Paper*, Number 6, November. http://www.biotech-info.net/Technical_Paper_6.pdf.

Burke, M. (2005) *Managing GM crops with herbicides: effects on farmland wildlife,* Produced by the Farm Scale Evaluation Research Committee and Scientific Steering Committee. http://www.defra.gov.uk/environment/gm/fse/results/fse-summary-05.pdf.

Cabinet Office Strategy Unit (2003) GM food in the late 1990s. In: Field Work: weighing up the costs and benefits of GM crops. Crown Copyright. http://www.strategy.gov.uk/downloads/su/gm/downloads/gm_crop_report.pdf.

Ching, L.L. and Matthews, J. (2001) GM crops failed. I-SIS. http://www.i-sis.org.uk/GMcropsfailed.php.

Department for the Environment, Food and Rural Affairs (DEFRA) (2002) GM crop farm-scale evaluations. Voluntary Agreement on the FSEs between government and the farming and biotechnology industry body (SCIMAC): Farm-scale evaluations of GM crops in the UK. http://www.defra.gov.uk/environment/gm/fse/scimac/agreement.htm.

DEFRA (2004a) GM (Genetic Modification): introduction to Genetically Modified Organisms (GMOs). Page last modified 15 July, 2004. http://www.defra.gov.uk/environment/gm/background/index.htm.

DEFRA (2004b) GM Dialogue. Page last modified 15 July 2004. http://www.defra.gov.uk/environment/gm/debate/index.htm.

DEFRA (2005) GM (Genetic Modification): answers to some frequently asked questions about GM crops. Updated 19 March 2005. http://www.defra.gov.uk/environment/gm/crops/faq.htm.

DEFRA (2005b) The farm scale evaluations. http://www.defra.gov.uk/environment/gm/fse/index.htm.

Department of Trade and Industry (DTI) (2003) *GM NATION? The findings of the public debate.* Crown Copyright. http://www.gmnation.org.uk/docs/gmnation_finalreport.pdf.

Fernandez-Cornejo, J. and McBride. W.D. (2002) Adoption of bioengineered crops, *Agricultural Economic Report. Economics Research Service, United States Department of Agriculture,* No. AER810. http://www.ers.usda.gov/publications/aer810/.

Firbank, L.G., Perry, J.N., Squire G.R., Bohan, D.A., Brooks, D.R., Champion, G.T., Clark, S.J., Daniels, R.E., Dewar, A.M., Haughton, A.J., Hawes, C., Heard, M.S., Hill, M.O., May, M.J., Osborne, J.L., Rothery, P., Roy, D.B., Scott and R.J., Woiwod, I.P. (2003) The implications of spring-sown genetically modified herbicide-tolerant crops for farmland biodiversity: a commentary on the farm scale evaluations of spring-sown crops. http://www.defra.gov.uk/environment/gm/fse/results/fse-commentary.pdf.

Firbank, L.G., Rothery, P., May, M.J., Clark, S.J., Scott, R.J., Stuart, R.C., Boffey, C.W.H., Brooks, D.R., Champion, G.T., Haughton, A.J., Hawes, C., Heard, M.S., Dewar, A.M., Perry, J.N. and Squire, G.R. (2005) Effects of genetically modified herbicide-tolerant cropping systems on weed seedbanks in two years of following crops. *Biology Letters.* http://www.journals.royalsoc.ac.uk/openurl.asp?genre=article&id=doi:10.1098/rsbl.2005.0390.

Fuller, R.J., Norton, L.R., Feber, R.E., Johnson, P.J., Chamberlain, D.E., Joys, A.C., Mathews, F., Stuart, R.C., Townsend, M.C., Manley, W.J., Wolfe, M.S.,

Macdonald, D.W. and Firbank, L.G. (2005) Benefits of organic farming to biodiversity vary among taxa, *Biology Letters*. http://www.journals.royalsoc. ac.uk/openurl.asp?genre=article&id=doi:10.1098/rsbl.2005.0357.

GM Science Review Panel (2003) *GM Science Review First Report: An open review of the science relevant to GM crops and food – based on interests and concerns of the public*. London: HMSO. http://www.gmsciencedebate.org.uk/report/pdf/ gmsci-report1-full.pdf.

GM Science Review Panel (2004) *GM Science Review Second Report: An open review of the science relevant to GM crops and food based on interests and concerns of the public*. London: HMSO. http://www.gmsciencedebate.org.uk/ report/pdf/gmsci-report2-full.pdf.

GM Watch (2004) Major problems in China with GM cotton. http://www.gmwatch. org/archive2.asp?arcid=3636.

Ho, M-W. (2005) Scientists confirm failures of Bt-Crops. I-SIS. http://www.i-sis. org.uk/SCFOBTC.php.

Hole, D.G., Perkins, A.J., Wilson J.D., Alexander, I.H., Gricee, P.V. and Evans, A.D. (2005) Does organic farming benefit biodiversity? *Biological Conservation*, 122 (1): 113–130.

Institute of Science in Society (I-SIS) (1999) Special safety concerns of transgenic agriculture and related issues. Briefing Paper for Minister of State for the Environment, The Rt Hon Michael Meacher. http://www.i-sis.org.uk/meacher99.php.

James, C. (2005) ISAAA Briefs 32-2004: Preview: Global status of commercialized Biotech/GM Crops: 2004. http://www.isaaa.org/kc/CBTNews/press_release/ briefs32/ESummary/Executive%20Summary%20(English).pdf.

Losey, J.E., Rayor, L.S., and Carter, M.E. (1999) Transgenic pollen harms monarch larvae. *Nature*, 399: 214.

MORI (2004) http://www.mori.com/polls/2003/uea-top.shtml.

Oberhauser, K.S., Prysby, M.D., Mattila, H.R., Stanley-Horn, D.E., Sears, M.K., Dively, G.P., Olson, E., Pleasants, J.M., Lam, Wai-ki F., and Hellmich, R. (2001) Temporal and spatial overlap between monarch larvae and corn pollen. *PNAS Early Edition*. www.pnas.org/cgi/doi/10.1073/pnas.211234298.

P G Economics Ltd. (2003) Consultancy support for the analysis of the impact of GM crops on UK farm profitability. http://www.pgeconomics.co.uk/pdf/Cabinet_ Office_GM_Crops.pdf.

Plant Managers Network (2002) Glyphosate-resistant waterhemp moves into the Corn Belt. 13 December 2002. *Plant Health Progress*. http://www. plantmanagementnetwork.org/pub/php/news/waterhemp/.

Pleasants, J.M., Hellmich, R.L., Dively, G.P., Sears, M.K., Stanley-Horn, D.E., Mattila, H.R., Foster, J.E, Clark, T.L., and Jones, G.D. (2001) Corn pollen deposition on milkweeds in and near cornfields. *PNAS Early Edition*. www.pnas. org/cgi/doi/10.1073/pnas.211287498.

Sainsburys (2005) http://www.j-sainsbury.co.uk/files/reports/cr2005/?pageid=52.

Sears, M.K., Hellmich, R.L., Stanley-Horn, E., Oberhauser, K., Pleasants, J.M., Mattila, H.R., Siegfried, B.D. and Dively, G.P. (2001) Impact of Bt corn pollen on monarch butterfly populations: A risk assessment. *PNAS Early Edition*. www. pnas.org/cgi/doi/10.1073/pnas.211329998.

Soil Association (2003) Flavr Savr tomato and GM tomato puree: The failure of the first GM foods. *Soil Association Briefing Paper*, 11/06/2003. http://www.soilassociation. org/web/sa/saweb.nsf/0/80256cad0046ee0c80256d1f005b0ce5?OpenDocument.

Stanley-Horn, D.E., Dively, G.P., Hellmich, R.L., Mattla, H.R., Sears, M.K., Rose, R., Jesse, C.H., Losey, J.E., Obrycki, J.J., and Lewis, L. (2001) Assessing the impact of Cry1Ab-expressing corn pollen on monarch butterfly larvae in field studies. *PNAS Early Edition*. www.pnas.org/cgi/doi/10.1073/pnas.211277798.

Tesco (2005) http://www.tescocorporate.com/page.aspx?pointerid=2C8F604AACC5 4868963C4121B14294BD&faqelementid=75A6071993684A77B75C17ED114 29F15.

United States Census Bureau (2005) http://www.census.gov/ipc/www/worldpop.html

United States Department of Agriculture (USDA) (2005) *Agricultural Chemical Usage 2004 Field Crops Summary*. May 2005. http://usda.mannlib.cornell.edu/ reports/nassr/other/pcu-bb/agcs0505.pdf.

Zangerl, A.R., McKenna, D., Wraight, C.L., Carrol, M., Ficarello, P., Warner, R., and Berenbaum, M.R. (2001) Effects of exposure to event 176 Bacillus thuringiensis corn pollen on monarch and black swallowtail caterpillars under field conditions. *PNAS Early Edition*. www.pnas.org/cgi/doi/10.1073/pnas.171315698.

Chapter 7

Balancing Nature Conservation 'Needs' and Those of Other Land Uses in a Multi-Functional Context: High-Value Nature Conservation Sites in Lowland England

Christopher Short

Introduction

This chapter focuses on areas of high-value nature conservation that are also multifunctional sites and highlights the different approach needed for their effective management. Such areas of land have relatively weak ties to agriculture but are increasingly important for other land-based interests such as nature conservation, recreation, heritage and landscape. The chapter looks at reasons behind the gradual detachment from agriculture of these parcels of land and how this impacts on productivst and post-productivist discussions within the rural studies literature (Evans *et al.*, 2002; Holmes, 2006; Wilson, 2001). The example of lowland England suggests that as productivist agriculture retreats it creates pockets of post-productivism at the local scale. It is argued here that the more agriculturally marginal these areas of land become the more multifunctional they are and the more important they are for other land-based interests. Thus heightened protection and consumption interests replace the production interests creating a triangular dynamic system, as outlined by Holmes (2006). The most pronounced example of this type of land is common land, and the nature and extent of this type of land is also outlined and explained together with the framework for a different approach to nature conservation.

As agriculture declines in overall economic terms, areas of intensive production seem to retreat while other land uses increase in both importance and influence. One such land use is nature conservation and the number of areas of significant importance for nature conservation is growing. However, ownership and thus day-to-day management responsibility often remains with the private individual. Consequently government officers or conservation NGOs increasingly need to work alongside other land uses in developing strategies covering nature conservation and other land uses in such places. Herein lies the dilemma for nature conservationists, on multi-functional sites where they may have some legal power to enhance the nature conservation management but this needs to be balanced with the importance of these sites to other users, making the management of these sites more complex. The nature conservationists' ability to have a dialogue with other users is crucial and tests their ability to see sustainability as a process of consensus building rather than optimum nature conservation management. This chapter will look at the discourse associated with this process and the techniques used by conservationists to obtain their 'goal'. The development of techniques such as conservation grazing, which seeks to imitate extensive agricultural grazing but without the need to link into the food production perspective (Oates and Tolhurst, 2000), is examined.

Common land is an excellent example in which to examine these issues for three main reasons. First it is the most frequent example of high-value conservation sites within lowland England. Second, there is legal protection for a range of other land uses making the management of these multifunctional spaces significantly more 'public' than in other areas. Third, the variety of other land-based interests is pronounced due to the lack of soil disturbance and agricultural improvement (Short and Winter, 1999). It is on such sites that the approach and the science of nature conservation are questioned. The chapter concludes by suggesting that the best approach might be a type of 'accountable conservation', which might have long-term benefits for nature conservation. However, since the 'practice' is ahead of the 'proof', further research into both the governance processes and management approaches is required.

The Recent Past

There is universal consensus that the period between the 1970s and into the 1990s is one where rural policy was driven by what was happening within the agricultural sector. Potter and Lobley (1998) outline the factors underpinning the dominance of agricultural policy in shaping the landscapes of Europe, based largely on the premise that policy and action would in reality follow the money. Since during this period the Common Agricultural Policy (CAP) brought considerably more money into the rural economy than any other official source, other interests were essentially little more than sideshows. The key driver for the CAP was more efficient production so this period became known as the productivist period (Winter, 1999).

During the nineties there were changes in the CAP regime, such as the 'MacSharry Reforms' in 1992, but these essentially tinkered with the production regime and provided guidance and grants based around efficient production (Winter *et al.*, 1998). Across Europe agri-environment schemes were developed in the late 1980s and grew in importance over the next decade. However, these schemes remained a voluntary land management option available to farmers but limited to either specific areas (e.g. the Environmentally Sensitive Areas [ESAs] scheme) or in terms of overall budget (e.g. the Countryside Stewardship Scheme [CSS]) and thus a minor role compared to conventional agriculture (Buller *et al.*, 2000; Potter and Lobley, 1999). In 2000 with the adoption of the Rural Development Regulation (RDR) agri-environment schemes across Europe received a major cash injection. In England the introduction of Environmental Stewardship has extended the accessibility of the ESA scheme to all farmers (Entry-Level Scheme) and the enhancement focus of CSS to the Higher Level Scheme (DEFRA, 2006). Across Europe there is an increasing realisation that, while agriculture remained responsible for the majority of land management and cultivation, it was rapidly declining as an economic influence. As a result the European Union (EU) has responded to changing rural circumstances and to a rural economy that is increasingly less orientated around agriculture by broadening the CAP towards a multi-sector rural development approach (van der Ploeg *et al.*, 2000).

The debates surrounding the presence or absence of post-productivism continue (Evans *et al.*, 2002; Mather *et al.*, 2006; Wilson, 2001), but most academics and analysts accept that productivism has not disappeared entirely, undermining the idea of 'post'. Indeed, one of the three axes that forms the core of the new RDR regulation is based around the 'competitiveness of agriculture'. Ilbery and Bowler (1998) coined the phrase 'the majority of the food from a minority of the land', which seems to be holding true as parts of the UK, especially the South-East, are becoming a focus for intensive levels of production. In this sense agricultural productivism might be seen as being in retreat rather than in wholesale decline.

In UK policy-terms the aim has been to develop policy along a twin-track approach embracing both land and enterprise (DEFRA, 2004). Across Europe it is acknowledged that two key features of this paradigm shift are increased diversity and multifunctionality (Knickel and Tenting, 2000). The issue of diversity relates to the range of actors involved in rural issues and the acknowledgement of multifunctionality concerns the 'simultaneous and interrelated provision of different functions' (van der Ploeg *et al.*, 2000). There are some suggestions that agriculture itself is multifunctional (Wilson, 2001), but as Potter and Burney (2002) point out this is not really novel or indeed controversial. Grant (2004) goes on to suggest that agriculture as an industry is developing certain aspects, such as organic production, as a more acceptable face while 'productivist' agriculture continues elsewhere. In marginal areas, agriculture and other land uses are increasingly becoming equals. However, to suggest that the countryside is somehow homogenous or that the transition is taking place equally across the United Kingdom is clearly not accurate, as the next section explores.

Pockets of Post-Productivism

The geographical focus of this chapter is lowland England, an area of high agricultural productivity. Within the more extreme productivist areas there are marginal areas that have largely been abandoned by conventional agriculture. These areas of land have increasingly weak ties to agriculture but are now progressively more important for nature conservation, recreation, heritage and landscape. It seems that the more agriculturally marginal these areas of land become the more multifunctional they are and the more important they are for non-agricultural land uses creating pockets of post-productivism. Such regional variation and local-scale analysis has been lacking from the discussion surrounding post-productivism. However, it could be accommodated within the 'multifunctional rural transition' outlined by Holmes (2006). Here the rather unhelpful continuum of productivism to post-productivism is replaced by a triangular approach based around the core purposes for rural land use:

namely, production, consumption and protection (see Figure 7.1). Holmes (2006: 143) suggests that the dynamic created by the tension between the 'agricultural overcapacity' (production), 'market driven amenity uses' (consumption) and 'changing societal values' (protection) all contribute to multifunctionality and signify the range of goals that different interests are seeking to achieve. Therefore, within this framework it is possible to understand the social, economic and environmental processes that shape the countryside and to highlight how different areas adjust and accommodate various combinations of land uses while others do so to a lesser extent.

Issues concerning nature conservation would normally be placed under the protection part of the triangular framework, but as a land use it is only sometimes the single land use. An example might be the management of National Nature Reserves (NNR), which is undertaken by Natural England, the successor to English Nature. The NNR promise states that these are 'key places for wildlife and natural features ...' and 'managed, on behalf of the nation, to maintain their special wildlife and natural features'. Crucially the promise goes on to state 'NNRs are places where wildlife and natural features come first' (English Nature, 2006). Within the United Kingdom the most significant designation in terms of land area is a Site of Special Scientific Interest (SSSI), also established under the Wildlife and Countryside Act 1981. However, the designation does not provide a legal basis for management, and those responsible, often government officers or conservation NGOs, need to work alongside other land uses in developing strategies for such places. In the majority of cases this means nature conservation working alongside the agricultural managers

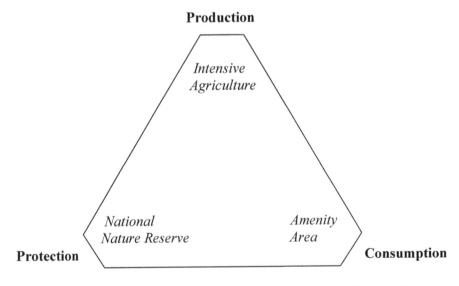

Figure 7.1   A Multifunctional Triangulation (adapted from Holmes, 2006)

adjusting the management to enhance biodiversity. Thus the production and protection aspects coexist in a management sense as indicated by type B in Figure 7.2.

This chapter focuses on a third type of nature conservation management (Type C), those sites where there are significant levels of consumption and other types of protection. In terms of consumption these sites may be close to a local community who have a high affinity with this area of local space. There may also be local land-based businesses, such as riding stables, which are significant users of the area. It is likely that such areas of low agricultural use will have been designated as 'open country' and so have a legal right of open access under the Countryside and Rights of Way (CRoW) Act 2000. Landscape perspectives may also be important through a local designation, such as the local planning designation of 'High Landscape Value' or the site may be within an Area of Outstanding Natural Beauty (AONB). In terms of heritage the sites may be recorded under the Sites and Monument Record (SMR) held by the local authority or designated as a Scheduled Monument (SM) by English Heritage (EH). In the case of the latter any soil disturbance within the designated area would require the consent of EH. In Type C cases the production function is a distant third compared to matters of protection and consumption as indicated in Figure 7.2. However, it is the production function that established the value of the site in the first place, and therefore some sort of imitating land management approach is required that replicates the production function.

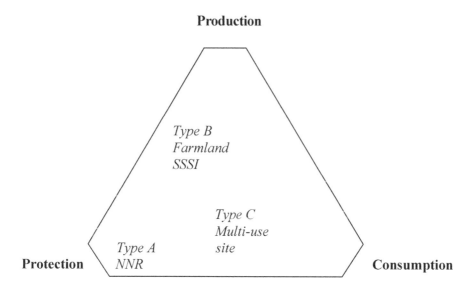

Figure 7.2   Types of Nature Conservation Management (adapted from Holmes, 2006)

The heightened awareness of nature conservation interests to achieve the establishment of good conservation management on SSSIs is related to a Public Service Agreement (PSA) established in 2000 as part of the CRoW Act 2000. It is worth noting that the inclusion of a significant section concerning biodiversity within the CRoW Act 2000 indicates a marked increase in both interest and action surrounding biodiversity since the signing of the Convention on Biological Diversity of Species and Habitats at the 1992 Rio Summit. These PSA targets are established by agencies in return for increased resources from government. The target agreed by English Nature was for 95 per cent of all SSSIs to be in favourable condition by 2010. By the end of 2005 69.8 per cent of SSSIs were in favourable or favourable recovering condition (English Nature, 2006). Of the remaining 33 per cent of SSSIs that are not in favourable condition over a quarter are registered common land. In the case of some habitats, such as lowland heath, which is concentrated in areas such as the South-East, this is considerably higher. The next section offers a background to common land generally and outlines why it is the archetype Type C example of conservation management.

Why is Common Land Different?

Common land is a pronounced example of multifunctional land use quite simply because it is valued for a number of different organisations for a wide range of reasons. The overall difference is in the type and level of management common land has received and the legislation developed to protect common land. The existence of extensive traditional management over long periods has also added considerably to the nature conservation value of common land and ensured that the land is characteristically different from other land around it.

Figure 7.3 shows that most common land is found in the upland areas of the north and west of England. However it also shows that a significant scattering of smaller commons still remain in the south-east part of England and that common land is largely absent from the midlands. This should not be a surprise as common land's historical origins are based around an assumption that the land was marginal and unproductive (Hoskins and Stamp, 1963). This early labelling, together with the 'shared' nature of these areas, resulted in commons being largely excluded from the effects of the agricultural revolution and the productivist policies that have been implemented in the UK over the past 50 years. Short and Winter (1999) argue that commons have experienced a type of 'constrained productivism' and therefore did not experience the intensification in management that other types of land, such as owner-occupied land did. As a result of these complex legal characteristics common land areas were left relatively untouched, although they are still managed through the grazing of the commoners' stock. The by-product of not being fully included within the drive for increased food production is their increasing importance for the range of interests outlined above.

In lowland areas where land could be managed more intensively, areas of marginal land were gradually left with little or no management. Alongside the decline of agriculture in overall land use terms, the rural economy was growing and there was

Figure 7.3   Registered Common Land in England and Wales

Source: DEFRA, 2006.

an increasing willingness to recognise other land uses in lowland rural areas. Many of these land uses, such as recreation, heritage and nature conservation, are now very important to common land and reflect the increased level of interest in such activities by local people and visitors, as well as membership organisations, such as the National Trust where 25 per cent of the land the Trust manages is common land. Given the formation of Natural England, which brings together the landscape, recreation and access functions of the Countryside Agency, the nature conservation responsibilities of English Nature and the Rural Development Service of DEFRA, it is not surprising that they have identified the sustainable management of commons as a 'priority area' (Bathe, 2006).

While the upland areas of south and north-west England and mid and north Wales are very significant for upland habitats, the sprinkling of green in the southern parts suggests a large number of commons in lowland parts of England. For example, common land makes up 31 per cent of lowland heath SSSIs, the habitat with the lowest favourable condition figure in England and the UK as a whole (English Nature, 2003; 2006). Nationally, over half of all commons are designated as SSSIs because of their importance to biodiversity and geology, covering a wide range of habitats from salt marshes to flower-rich grassland as well as hill and mountain grazing areas.

For many local communities 'the common' is a key part of their daily lives, providing fresh air, a sense of belonging and identity, and room to stretch on untamed and open land that is very different from the surrounding landscape. In the past common land has been fundamental to traditional rural existence for centuries, and remains vital to the economy of many areas. Common land is rich with archaeological information that has often been lost elsewhere because the majority of land has been cultivated and the time-line disturbed. The low level of management and minimal soil disturbance compared with other areas of land has heightened the importance of common land as a 'time capsule of past activity' (Short and Winter, 1999). According to Rackham (1986) common land was the focus for many early settlements and so regularly hosts barrows, tumuli and hill forts, reflecting the local characteristics of the area (Hooke, 1993). For example, one lowland common in South Wales contains two Neolithic chambered tombs, 12 burnt mounds, 12 structured cairns, 14 large cairns and over 200 smaller cairns (Ward, 1989).

Landscapes are places where 'people and nature meet' and interact with one another (Brown *et al.*, 2005), while for others landscapes are important indicators of cultural and regional identity (Knickel and Renting, 2000) that reflect the basic organisation of society and economy (Terkenli, 2001). Half of the commons within England lie within National Parks and a further 31 per cent are in AONBs, areas largely in the southern part of the country, designated for their landscape value (English Nature, 2006).

Evidence for recreation and access also suggests that many commons receive hundreds of thousands visitors each year and are very important areas of local open space (Short and Winter, 1999). The CRoW Act 2000 introduced a legal right of access (*de jure*) for 'walking and quiet enjoyment' on large areas of 'open country', including all registered common land, and has clarified a previously complex situation. Figure 7.4 shows the location of the 947,000 hectares of open access land in England (7 per cent of the land area), of which 38 per cent is common land including significant areas in the lowlands. It is worth noting that the occurrence of open access land mirrors that of common land, with large parts of the midlands largely empty while southern England provides a scattering of open access land alongside the larger tracts within the upland areas of both England and Wales. Following these changes nearly all common land in England is now open to the public under a legal right of access rather than a permissive one (English Nature, 2003). The only exceptions are small areas where exclusions have been agreed through government agencies, possibly because of military training or especially sensitive areas of biodiversity.

A further and more profound difference concerns the legislation that relates specifically to common land. The legislation is complex and relevant Acts span two

Figure 7.4   Open Access Land in England and Wales

Source: Bathe, 2005.

centuries or more, but its central theme is to keep the land from both 'enclosure' and development (Short and Winter, 1999). The process of enclosure began in earnest during the late sixteenth and early seventeenth centuries, largely focussing around areas of high population expansion (Rackham, 1986). The passing of the Inclosure Act 1845 was meant to ensure total enclosure but this proved a step too far and after much protest the proposed Act was withdrawn (Short, 2000). This u-turn was complete when the Metropolitan Commons Act of 1866 was passed, which protected many commons, notably those around London, as 'green lungs' for the whole population (Clayden, 2003). Since that date there have been a number of further pieces of legislation that ensure the special status of common land, all of which contributes to its complex, and often misunderstood, character. For example, under the Law of Property Act 1925 permission is required from the Secretary of State for any works on commons such as the erecting of fences. Enclosure seems a dated idea in the early part of the 21st century, but it might translate more effectively as saying that the legislation is in place to prevent common land from becoming like other types of land and being the responsibility of one, rather than a number of, individuals, organisations or, indeed, land uses. There have been various attempts to rationalise this legal minefield (Common Land Forum, 1985; Royal Commission, 1958) and others to translate it (Clayden, 2003; DETR, 1998; English Nature, 2000; Gadsden, 1988). Most recently the Government gained royal assent for the Commons Act 2006 that seeks to:

- enable the establishment of statutory commons associations with powers to sustainably manage commons by majority decision making;
- prevent severance of common rights from the land to which they are attached;
- modernise registration of common land and common rights, including allowing missed commons to be properly recognised, and taking wrongly registered land off the register; and
- reinforce existing protections against abuse, encroachment and unauthorised development.

Investment in Non-Agricultural Interests

As indicated in a previous section, funding allocated to environmental enhancement under the CAP has increased significantly. In addition there have been other investments through government, linked to the PSA target, and non-governmental sources. For example, the Heritage Lottery Fund (HLF), which has provided £67.5 million into conservation projects over the past five years (Wildlife Trusts, 2004). As with any initiative, with this amount of investment over a relatively short space of time changes needed to be identified in order to attract the investment. Overall, £24.5 million has been directed towards 183 land acquisitions with £29.4 million allocated for 'capital works' on Wildlife Trust (WT) reserves, including improved visitor access, volunteer training machinery and specific features such as hides as well as fencing for grazing stock (Wildlife Trusts, 2004). This is in addition to the significant investment arising out of the CAP reforms where monies within the pillar 2 'rural development' funds have led to a doubling of the funds provided in this areas and a growth in not only land-based agri-environment schemes, such as Environmental Stewardship, but also project-based environmental schemes, such as the Rural Enterprise Scheme.

There has been funding for other areas too, towards similar sized initiatives in community and heritage initiatives. A good example is the Local Heritage Initiative operated by the Countryside Agency and also funded by the HLF with the aim of providing grants to local groups and communities to enable them to 'celebrate their heritage' by helping them to 'investigate, explain and care for their local landscape, landmarks, traditions and culture' (Local Heritage Initiative, 2004, online). Access and recreation have experienced considerable growth generally and there has been a huge policy investment through the implementation of the CRoW Act 2000.

All of this investment increases the expectations of the recipient interest group and a subsequent 'delivery' of the established objectives. Clearly this is most straightforward where there is a single issue or area of control, as in the type A example, a nature reserve owned by a conservation body, where it would be relatively straightforward to implement management changes over a five-year period. Even on type B areas it is possible through agri-environment schemes to introduce the required changes to enhance the nature conservation interests. However, in cases of Type C conservation management where there are a larger number of stakeholders, there is a greater likelihood that frustration and disagreement can occur. Given the range of interests involved and the complex legal situation, commons are a good example of situations where the views and management objectives of all stakeholders need to be considered. Frustration and disagreement are often increased when the achievement of a particular interest's objectives are delayed because of objections from other stake-holders or because of the need to consult other stakeholders for some management decisions as required by the legal processes relating to common land.

Therefore multifunctional land, such as common land, requires a different approach to nature conservation management. The next section outlines the approaches being adopted and suggests that in the long-term this is ultimately a more sustainable type of nature conservation because other interests are educated about the needs and aspirations relating to biodiversity. Nature conservation itself also becomes more accountable for its actions especially in cases where there are changes in existing management or a new overall management strategy is developed or introduced.

Accountable Nature Conservation

It is widely accepted now that sustainability is divided equally between environmental, economic and social aspects (Pierce, 1996) and these are increasingly finding their way into nature conservation strategies (Adams, 2003). Consequently, there are a number of examples where issues relating not only to landscape and heritage but also people and the local economy are considered alongside nature conservation matters in order to be effective and sustainable in the long-term (Bishop and Phillips,

2004). In this respect the achievement of sustainable management on multifunctional land, such as common land, is no different. An earlier part of this chapter has already shown that a large proportion of multifunctional land, like common land, is valued under a range of nature conservation designations, both nationally and internationally. This high environmental value has been achieved through traditional 'custom and practice' and a system based on trust and cooperation, albeit through extensive agricultural practices.

Changes within agriculture and in rural land use more generally have placed these systems under considerable strain and in some areas such practices may be part of the folk law. It is clear from the research into common land (Short and Winter, 1998; Short *et al.*, 2005) that reviving and encouraging a struggling existing system is far more likely to be effective than restarting one that has ceased altogether or replacing it with a system geared to a single objective. Given the wide variety of interests involved in common land and other types of multifunctional land, it is also clear that the more recent developments within nature conservation recognise the importance of cultural, historic and socio-economic aspects in the landscape. A good example would be the Lifescapes initiative involving English Nature, English Heritage and the Countryside Agency, where the Natural Areas and Countryside Character provide a way forward for a more integrated view of landscape and management planning (Porter, 2004). This often requires a different set of 'tools', both mechanical and animal to do the work; here the effort of the Forum for the Advancement of Conservation Techniques (FACT) and the Grazing Animals Project (GAP) are crucial in 'thinking outside the box' in terms of nature conservation approaches. However, such approaches often end up replicating the 'custom and practice' through the use of nature conservation expertise. This is the case where conservation organisations buy in and circulate stock to graze a number of areas experiencing under-grazing. A much more sustainable approach has been developed in the Cotswolds where a grassland strategy has been developed that promotes engagement with the local community, especially farmers neighbouring the land in question. The result has been the adaptation of 'custom and practice' through the use of local stock with the conservation organisation providing advice and support (Cotswold Conservation Board, 2004).

Increasingly, including community participation within nature conservation is replacing the fragmented sectoral approach to issues in Britain. This is perhaps most pronounced in Scotland and Wales where, following devolution, the areas of countryside and conservation have been brought together in a single agency. For Scotland it is Scottish Natural Heritage and for Wales the Countryside Council for Wales. The resulting programmes, such as the Futures Programme in Scotland that seeks an active engagement with communities regarding the management of protected sites, seem to suggest a greater commitment to the integrated approach (Crofts, 2004). Although this is but one part of a more integrated jigsaw, the formation of Natural England might be seen as a step in the right direction as would the indication that Natural England might become a 'champion' for common land (Bathe, 2006).

From an economic perspective the value of common land is difficult to determine, not least because of the number of variables. In market economic terms much of common land is extremely marginal, especially in lowland areas, but what is clear is that common land has a 'value' for many interests, ironically largely because it was previously judged to be marginal from an agricultural perspective. Therefore, making these areas economically viable again in the traditional sense is not likely to be successful. Nevertheless it is important to replicate the traditional customs and practices as far as possible, and through agri-environment schemes existing graziers can be encouraged to continue if the payments are sufficient (Short, 2000). It is also possible these areas can provide existing commoners with real opportunities to use the 'value' of the common to 'add value' to the products derived from the stock that graze these areas. For an example see the website of *Forest Friendly Farming* (Rigglesworth, 2005). In cases where there is no farmer involvement conservation agencies sometimes manage stock and graze areas; this is called conservation grazing (Oates and Tolhurst, 2000). There are examples of conservation grazing where, with sufficient support and investment of considerable time, it is possible to generate some financial resources that can be fed back into other areas of management but to call such schemes 'profitable' is a little wide of the mark (Oates and Tolhurst, 2000). The economics of grazing for conservation need to be clearly established and this should include time for stock management and animal welfare issues as well as the need for back-up land should the stock need to be removed from the grazing area (Oates and Tolhurst, 2000). A crucial part of the economic aspect of sustainability is the need to consider the long-term nature of any proposal. For this reason locally-based solutions that are not dependent on project staff or stock travelling a large distance to graze the site are more sustainable (Cotswold Conservation Board, 2004).

The social aspect of sustainability on common land sites is determined by their local distinctiveness and the richness of the history. The recent findings of a survey for the National Trust (MORI, 2004) show just how important 'local' space is to those who live in this area. The historical connection of communities to common land has often been explored by local societies or, if it has not been, would make an excellent opportunity for community involvement. The initial key sources would be the Sites and Monument Record (SMR) held by the local authority, the Countryside Character approach established by the Countryside Commission (now the Countryside Agency) (Swanwick and LUC, 2002) and evidence of involvement in the historic landscape characterisation process by the local authority or others (Macinnes, 2004).

In many cases the types of knowledge that are being employed pitch the scientific against local, or lay, knowledge. There is a significant body of literature to suggest that this is not an easy relationship (Clark and Murdoch, 1997: Wilson and Hart, 2001). There are plenty of examples where the prescriptions associated with nature conservation schemes have been imposed from a national level under the disguise of scientific knowledge and replacing local custom and practice (Short, 2000; Mills *et al.*, 2006). There is increasing evidence that the traditional approach to such situations of using participation and consultation techniques are unsuitable in situations like common land where there are 'high degrees of interdependency, complexity, uncertainty and multiple stakeholding' (Collins and Ison, 2006: 2). They go on to highlight the difference between traditional policy approaches where there are 'fixed forms of knowledge' and social learning where 'the knowing occurs during the process' (Collins and Ison, 2006: 10).

Conclusion – A Way Forward?

The management of multifunctional sites is complex, testing the nature conservationists' ability to have a dialogue with other users and their ability to see sustainability as a 'process' of building greater understanding and knowledge for all parties rather than optimum nature conservation management. Research commissioned by five organisations,[1] all concerned with the long-term future of common land (Short *et al.*, 2005), notes that in some cases the medium- to long-term future of nature conservation in multifunctional contexts is uncertain, as they are not receiving the type of management needed to maintain their current value. In an attempt to rectify this situation the resulting guide introduces and outlines a process concerned with the long-term management of multifunctional land such as common land. It essentially builds on the types of holistic processes that are beginning to become more widely accepted (Acland, 1992; Etchell, 1996) and developed further (Collins and Ison, 2006) within nature conservation. This focus on the process rather than the outcome fits common land very well because of the wide local variation and the fact that on any one area of common there will always be a number of interests and associated stakeholders, who in turn will have varying levels of knowledge and interdependency. Clearly the precise nature of the process, in terms of who initiates and drives it, the length of time it lasts and the level of detail involved, will vary from one location to another. For those commons where there is a history of widespread stakeholder involvement in the management planning process, some of the stages may be condensed. Nevertheless there are some 'golden rules' (see Figure 7.5) to underpin the aim of the guide for all users; they also form a basic starting point.

The approach is based on the principle of working with stakeholders, increasing levels of awareness of the various issues and the mutual identification of the range of possible solutions that would resolve these issues. In this sense it requires nature conservation to embrace people, as well as landscape, history and access issues as Adams (2003) predicted. Where this approach is taken there is anecdotal evidence that this provides an opportunity for people to reconnect with their local surroundings and with nature in particular (Short, 2006). As in social learning, the process of identifying the nature of the issue itself also highlights how a solution to the issue might be progressed. It is therefore important not to see participation as something linear as this 'fails to capture the dynamic and evolutionary nature of user involvement' (Tritter and McCallum, 2006: 165).

In parallel to Marsden's (1999) suggestion that agriculture 're-integrate' itself with the local economy, these multifunctional sites offer nature conservation, as well as other land uses, the opportunity of fully integrating themselves with other land uses. However, this does not sit well with the extensive experience that nature conservationists have with type A and B conservation sites. Given the time it has taken nature conservation to be embraced alongside agriculture (see Winter, 1999) it can only be hoped that this type of consensus-based management on areas of important but agriculturally marginal land are taken seriously, as the alternative might be some variant of 'productivist conservation' where other interests are side-lined in order to enhance the biodiversity present. It certainly seems to be the case that practice is ahead of proof regarding the type of conservation techniques that are

---

1. Multifunctional land is valued by many people for different reasons. What people value may differ but they are united by the strength of their concern.

2. Progress is least likely when one interest in the area attempts to sideline the others, or forces change upon them.

3. Regular communication amongst stakeholders is critical in building and maintaining trust and confidence between parties, and should start from a very early stage in the process.

4. Lasting progress is most likely when:

    - people respect and try to understand each others' values and aims;

    - people recognise that all perspectives are valid and that everyone will have things in common;

    - they keep an open mind about what form and change should take, until they have properly explored the various options and the impacts on others;

    - any change brings benefit to the neighbourhood and wider interests.

5. Complete unanimity may not be possible but a broad consensus should be the aim.

Figure 7.5   'Golden rules' of agreeing management on multifunctional land such as Common Land (adapted from Short *et al.*, 2005)

acceptable on multifunctional sites. There is a need for more research to 'test' both the processes suggested, as well as others, and the variations of management used on type A and B sites that is appropriate for Type C sites. Overall, this chapter has confirmed the heterogeneity of the English countryside at the micro-scale and the importance of retaining this level of diversity.

Note

1   Countryside Agency, English Nature, The National Trust, Open Spaces Society and the Rural Development Service (RDS) in the Department for Environment, Food and Rural Affairs (DEFRA).

References

Acland, A.F. (1992) Consensus-building: how to reach agreement by consensus in multi-party, multi-issue situations. London: The Environment Council.

Adams, W. (2003) *Future Nature*. London: Earthscan, second edition.

Bathe, G. (2005) *Natural England and common land*. Proceedings of National Seminar on Common Land and Village Greens, University of Gloucestershire, Cheltenham.

Bishop, K. and Phillips, A. (eds) (2004) *Countryside Planning: new approaches to management and conservation.* London: Earthscan.

Brown, J., Mitchell, N. and Beresford, M. (2005) Protected landscapes: a conservation approach that links nature, culture and community. In Brown, J., Mitchell, N. and Beresford, M. (eds) *The Protected Landscape Approach: linking nature, culture and community*, Gland, Switzerland and Cambridge: IUCN, pp. 3–18.

Buller, H., Wilson, G.A., and Höll, A. (eds) (2000) *Agri-environmental policy in Europe.* Aldershot: Ashgate.

Clark, J. and Murdoch, J. (1997) Local knowledge and the precarious extension of scientific networks: a reflection on three case studies. *Sociologia Ruralis* 37 (1): 38–62.

Clayden, P. (2003) *Our Common Land: The Law and History of Commons and Village Greens.* Henley-on-Thames: The Open Spaces Society.

Collins, K. and Ison, R. (2006) Dare we jump off Arnstein's Ladder? Social Learning as a new Policy Paradigm, *Proceedings of PATH Conference.* 4–7 June 2006, Edinburgh, Scotland. Available from: http://www.macaulay.ac.uk/PATHconference/PATHconference_proceeding_ps3.html.

Common Land Forum (1986) *The report of the Common Land Forum.* Cheltenham: Countryside Commission, Publication CCP215.

Cotswold Conservation Board (2004) *Grassland Strategy for the Cotswolds.* Northleach: Cotswold Conservation Board.

Crofts, R. (2004) Connecting the pieces: Scotland's integrated approach to the natural heritage. In Bishop, K. and Phillips, A. (eds), *Countryside planning: new approaches to management and conservation.* London: Earthscan, pp. 170–187.

Department of the Environment, Food and Rural Affairs (DEFRA) (2002) *Common Land Policy Statement.* London: DEFRA, July.

DEFRA (2003) *Consultation on agricultural use and management of commons.* London: DEFRA.

DEFRA (2004) *Rural Strategy 2004.* http://www.degra.gov.uk/rural/pdfs/strategy/rural strategy_2004.pdf.

DEFRA (2006) *Information on Common Land and Commons Legislation.* http://www.defra.gov.uk/wildlife-countryside/issues/common/index.htm (Accessed on 26 July 2006).

Department for Environment, Transport and the Regions (DETR) (1998) *Good practice guide on managing the use of Common Land.* London: DETR.

English Nature (2000) *Common land: unravelling the mysteries.* Peterborough: English Nature.

English Nature (2003) *Condition of SSSI habitats within common land in England.* Peterborough: English Nature.

English Nature (2006) *Statistics concerning SSSIs and BAP habitats.* Peterborough: English Nature.

Etchell, C. (ed.) (1996) *Consensus in the countryside: reaching shared agreement in policy, planning and management.* Proceedings from a Countryside Recreation Network (CRN) Workshop. Cardiff: CRN.

Evans, N.J., Morris, C. and Winter, M. (2002) Conceptualizing agriculture: a critique of post-productivism as the new orthodoxy. *Progress in Human Geography* 26: 313–332.

Gadsden, G. (1988) *The law of Commons*. London: Sweet & Maxwell.

Grant, W. (2004) *Is multifunctionality for real?* http://www.wyngrant.tripod.com/wyngrantCAPpage.html (Accessed on 6 October 2006).

Holmes, J. (2006) Impulses towards a multifunctional transition in rural Australia: Gaps in the research agenda. *Journal of Rural Studies*, 22 (2): 142–160.

Hooke, D. (1993) *Warwickshire's Historical Landscape – the Arden.* Birmingham: University of Birmingham.

Hoskins, W.G. and Stamp, L.D. (1963) *The Common Land of England and Wales*, London: Collins.

Knickel, K. and Renting, H. (2000) Methodological and conceptual issues in the study of multifunctionality and rural development. *Sociologia Ruralis*, 40(4): 512–528.

Ilbery, B.W. and. Bowler, I.R. (1998) From agricultural productivism to post-productivism. In Ilbery, B.W. (ed.) *The Geography of Rural Change*. Harlow: Longman, pp. 57–84.

Local Heritage Initiative (2006) *About Local Heritage Initiatives.* http://www.lhi.org.uk (Accessed on 6 October 2006).

Macinnes, L. (2004) Historic landscape characterisation. In: Bishop, K. and Phillips, A. (eds), *Countryside planning: new approaches to management and conservation.* London: Earthscan, pp. 155–169.

Marsden, T.K. (1999) Rural Futures: the consumption of the countryside and its regulation. *Sociologia Ruralis*, 39: 501–520.

Mather, A., Hill, G. and Nijnik, M. (2006) Post-productivism and rural land use: cul de sac or challenge for theorization? *Journal of Rural Studies*, 22: 441–455.

Mills, J., Gibbon, D., Dwyer, J., Short, C. and Ingram, J. (2006) *Identification of Delivery Mechanisms for Welsh Top-tier Agri-environment Schemes.* Confidential draft report to the Welsh Assembly Government, Cheltenham: Countryside and Community Research Unit.

MORI (2004) *Landscapes in Britain.* Research conducted for The National Trust. London: The National Trust.

Oates, M. and Tolhurst, S. (2000) Comment. Grazing for nature conservation: rising to the challenge. *British Wildlife*, 11: 248–353.

Pierce, J.T. (1996) The conservation challenge in sustaining rural environments. *Journal of Rural Studies*, 12: 215–229.

Potter, C. and Lobley, M. (1998) Landscape and Livelihoods: environmental protection and agricultural support in the wake of Agenda 2000. *Landscape Research*, 23 (3): 223–236.

Potter, C. and Lobley, M. (1999) Environmental Stewardship in UK Agriculture: a comparison of the Environmentally Sensitive Areas Programme and the Countryside Stewardship Scheme in South East England. *Geoforum*, 29(4): 413–432.

Porter, K. (2004) The natural area experience. In: Bishop, K. and Phillips, A. (eds), *Countryside planning: new approaches to management and conservation.* London: Earthscan, pp. 91–108.

Rigglesworth, E. (2005) Buy New Forest produce. The Forest Friendly Farming Leader+ Initiative. Lyndhurst: New Forest National Park Authority. www. forestfriendlyfarming.org.uk.

Rackham, O. (1986) *The History of the Countryside.* London: JM Dent.

Royal Commission on Common Land (1958) *Report of the Royal Commission on Common Land,* 1955–1958. London: HMSO.

Short, C. (2000) Common land and ELMS: a need for policy innovation in England and Wales. *Land Use Policy* 17: 121–133.

Short, C. (2006) *Future of Stroud Commons,* Report to the Cotswold Conservation Board and English Nature, Cheltenham: Countryside and Community Research Unit.

Short, C. with Winter, M. (1998) *Managing the use of Common Land.* Final research report to DETR. Cheltenham: Countryside and Community Research Unit.

Short, C. and Winter, M. (1999) The Problem of Common Land: towards stakeholder governance. *Journal of Environmental Planning and Management,* 42(5): 613–630.

Short, C., Hayes, L., Selman, P. and Wragg, A. (2005) *A common purpose: a guide to agreeing management on common land.* Report to the Countryside Agency, English Nature, The National Trust, Open Spaces Society and Rural Development Service, DEFRA. Peterborough: English Nature.

Swanwick, C. and Land Use Consultants (2002) *Landscape character assessment guidance.* Cheltenham: Countryside Agency, and Edinburgh: Scottish Natural Heritage.

Terkenli, T. (2001) Towards a theory of the landscape: the Aegean landscape as a cultural image. *Landscape and Urban Planning,* 57: 197–208.

Tritter, J.Q. and McCallum, A. (2006) The snakes and ladders of user involvement: moving beyond Arnstein. *Health Policy,* 76: 156–168.

Van der Ploeg, J., Renting, H., Brunori, G., Knickel, K., Mannion, J., Marsden, T.K., de Roest, K., Svilla-Guzmán, E. and Ventura, F. (2000) Rural development: from practices and policies towards theory. *Sociologia Ruralis,* 40(4): 391–408.

Ward, A. (1989) Cairns and "cairn fields"; evidence of early agriculture on Cefn Bryn, Gower, West Glamorgan. *Landscape History,* 11: 5–18.

Whitby, M. (2000) Challenges and options for the UK agri-environment: presidential address. *Journal of Agricultural Economics,* 51 (3): 317–332.

Wildlife Trusts (2004) Thanks a million. *Natural World,* Summer: 27–43.

Wilson, G.A. (2001) From Productivism to post-productivism and back again: exploring the (un)changed natural and environmental landscapes of European agriculture. *Transactions of the Institute of British Geographers,* 26: 77–102.

Wilson, G.A. and Hart, K. (2001) Farmer participation in agri-environmental schemes: towards conservation orientated thinking?' *Sociologia Ruralis,* 41(2): 254–274.

Winter, M. (1999) *Rural Politics: policies for agriculture, forestry and the environment.* London: Routledge.

Winter, M., Gaskell, P. with Gasson, R. and Short, C. (1998) *Effects of the 1992 CAP reforms.* Cheltenham: Countryside & Community Press.

PART 3
Sustainable Rural Communities

Chapter 8

'Culture Economy', 'Integrated Tourism' and 'Sustainable Rural Development': Evidence from Western Ireland

Mary Cawley and Desmond A. Gillmor

Introduction

A sustainable rural system is interpreted here with reference to 'sustainable development' as defined in the Bruntland Report (World Commission on Environment and Development, 1987); namely, development in such a way that productivity may be maintained over the longer term for future generations, whilst preserving essential natural systems and protecting human heritage and biodiversity. The European Commission (EC) and individual governments have now incorporated these principles of sustainability into policy, including rural development policy (Government of Ireland, 1997; CEC, 2001; EUROPA, 2003). Increasingly official rural development documents adopt a holistic approach to the definition of sustainability that involves environmental, economic, social, and cultural dimensions (usually in that order) but such policy documents are frequently lacking in the most effective methods of attaining their aims and may emphasise the technological and institutional to the neglect of the social and cultural (Jenkins, 2000). By contrast, Ray (1998a) highlights the opportunities offered to use cultural resources for economic purposes as a form of 'culture economy', drawing on the changing nature of post-industrial consumer capitalism, the trajectory of rural development policy in the European Union (EU) and the growth of regionalism. A culture economy approach may be described as capitalising on the distinctive features of local areas and cultural practices by commodifying them for commercial purposes rather than seeking to pursue scale economies in production. International development agencies suggest that such commodification or valorisation of often previously non-transacted commodities by local people should become an empowering experience, thereby increasing social capital (OECD, 1995; CEC, 1996a, 1996b). The alternative strategy of attracting external investment in search of profits may contribute to exploitative economic relationships and leakage from the local economy. This chapter focuses on the potential for pursuing a culture economy approach to the promotion of rural tourism as part of a broader rural development strategy. Kneafsey's (2001) research in the West of Ireland and Brittany provides evidence of such an approach in action. Bossevain (1996) and Butler et al. (1998) further illustrate that the scope for adding commercial value to cultural resources has been enhanced during the last decade by a growing demand for eco-tourism experiences and contact with traditional cultures.

A research project, conducted between 2001 and 2004 in six EU member states, including Ireland, on the theme 'Supporting and Promoting Integrated Tourism in Europe's Lagging Rural Regions' (SPRITE), proposed the concept of integrated tourism as a method of contributing to sustainable rural development (Jenkins and Oliver, 2001). Integrated tourism is grounded in culture economy, being defined as incorporating forms of tourism that are explicitly linked into the natural, economic and socio-cultural structures of the regions in which it takes place, and that seek to make optimal use of those structures. The objective of SPRITE was to analyse and develop the potential for better integration in tourism in the lagging rural regions of Europe. The project was innovative in seeking to develop a holistic definition of integrated tourism that included all relevant actors and instrumental dimensions that were perceived to facilitate the promotion of integration. These dimensions or nodes were conceptualized as relating to inherent features of tourism activity and the achievement of particular outcomes, which include multi-dimensional sustainability along the various structural dimensions and the empowerment of local people. Effective networking, at scales from the local to the international, was viewed as influencing the ways in which integration was promoted and the achievement of appropriate outcomes (Jenkins and Oliver, 2001).

In each of the six countries studied, one established and one less-established tourism region were selected so as to provide an internal comparative context for assessing the potential for promotion. All regions were defined as 'lagging' according to EU criteria. The West Region was selected as being the longer established tourism region in Ireland and is the focus of this chapter (a group of north midland counties was the less-established region). Six actor groups who are centrally involved in tourism were interviewed: business owners and managers (entrepreneurs), providers of resources for tourism, tourists, tour operators, host community members and institutions with roles pertinent to tourism. This chapter focuses on the entrepreneurs and the controllers of resources for tourism who are most closely engaged in the development of tourism infrastructure and products and their marketing, and whose support is central to the successful adoption of a culture economy approach. The purpose of the chapter is to examine the extent to which a culture economy pathway is in place in a lagging rural region with an established history of tourism and to assess the degree to which the future of tourism development, as perceived by centrally involved actors, aligns with such a pathway. The chapter begins by discussing the concept of 'culture economy' and the criteria on which integrated tourism was measured. The study method is then described, the results of the analysis are presented and conclusions are reached relating to the potential offered by a culture economy approach in promoting multi-dimensional rural sustainability and empowerment.

'Culture Economy' and 'Integrated Rural Tourism'

Convincing arguments have been advanced for incorporating cultural practices and environmental attributes within the discussion of economy in a growing body of geographical literature, at both a theoretical level and within particular sets of structural relationships, as a recent summary by Barnes (2005) illustrates. Such a hybrid development strategy is particularly appropriate in the lagging rural regions of Europe. In these areas negative structural features, such as mountainous topography and associated climatic limits on land use, small farms, distance from markets and ageing populations, have proven incapable of remediation in a satisfactory manner using conventional market economy approaches (Terluin and Post, 2000). At the same time these particular features of lagging rural regions, together with their cultural practices, are known to be attractive to tourists who seek contact with nature and with more traditional ways of life (Shaw and Williams, 2004). Herein lies an economic opportunity for local residents. An approach that integrates culture and economy involves the commodification of the intrinsic features of lagging rural areas to meet emerging demand as a method of supporting local economies. Cultures are thus viewed as sets of resources available for social and economic control (Ray, 1999). Ray (1998a) identified four different modes of culture economy emerging in the contemporary countryside, of which the first is of particular interest to the present study. In this mode, as noted by Lash and Urry (1994), cultures may be markers of territorial identity such as language, societal norms and traditions but also particularities of the physical environment that confer a locality with distinctiveness. These features may be imbued with quality attributes either informally, as when a producer establishes a reputation for providing a service of high standard, or formally through the award of an externally-monitored quality mark. Quality marks confer niche market recognition and command a price premium. The countryside per se may become commodified in novel ways as the medium through which recreational experiences are accessed, and quality designations for environmental features are emerging (for example, the EU Blue Flag award for beaches). Regional distinctiveness and the associated quality attributions may be constructed into a particular image for a locality and used both formally and informally in promotion and marketing (Ilbery and Kneafsey, 1998). Regional labelling pursued in a cohesive way constitutes the second mode of culture economy defined by Ray (1998a). Cultures may also be non-territorial, as reflected in the emergence of new philosophies relating to land use and methods of production that characterize the organic farming movement. Inherent in the pursuit of a culture economy approach, therefore, is the concept of the valorisation of local resources to meet new demands.

The invocation of cultural elements as part of a rural development strategy is facilitated by new forms of governance that have emerged during the last decade, based on partnership between local and extra-local public and private actors (Ray, 1998b; Sage, 2003). The support of both local and external agencies, as well as collaboration between local producers, is usually necessary to facilitate the valorisation process and to support promotion. Businesses in lagging rural regions tend to be small-scale privately owned enterprises with limited capital for investment, variable skill levels, and limited time and capacity to market their products (Skuras *et al.*, 2005). Official agencies have a role to play in helping them to overcome some of the deficiencies that exist and this role is recognized in both state and EU policy, including the LEADER programme (Ray, 1999). Moreover, in the case of rural tourism products external markets must be accessed because of limited local demand. Networking between producers and agencies therefore takes place at a horizontal local level and vertically from the local to the extra-local in what Murdoch (2000: 407) has described as 'a new paradigm of rural development'.

Culture Economy in the Context of Integrated Tourism

Integrated tourism was defined as being closely linked into the physical, economic, social and cultural environments in which it takes place through the use of local resources, the involvement of local actors, the mechanisms used to promote development, and the outcomes achieved. The linkages were conceptualised as relating to local ownership ('endogeneity'), 'embeddedness' in local structures, complementarity to other aspects of economy and society and the scale of tourism activities pursued (Figure 8.1). Multi-dimensional sustainability and the promotion of local empowerment were the anticipated outcomes. Networking between actors with vested interests in tourism across various geographical spaces from the local to the international was viewed as being instrumental to the achievement of integration (Jenkins and Oliver, 2001). The rationale for selecting the diagnostic criteria and the projected outcomes are now discussed.

'Endogeneity' relates to local ownership of resources and the sense of choice (local, collective agency) in how to employ those resources in the pursuit of local objectives (Ray, 1999). Such an approach should serve to retain economic benefits within an area. By contrast, external ownership of resources and external choice decisions remove that control from the local arena (Kemm and Martin-Quirós, 1996).

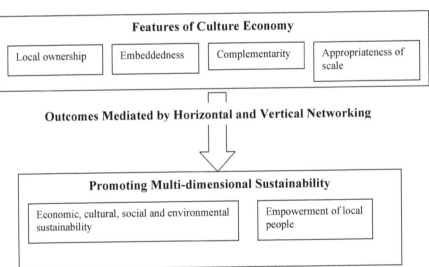

Figure 8.1   A Model of Integrated Tourism (based on Jenkins and Oliver, 2001)

It is important, therefore, that local entrepreneurship and investment are supported so as to increase the benefits for the locality. Related to endogeneity is the concept of 'embeddedness'. Embeddedness pertains to the extent to which tourism is linked into the local resource base; for example, the physical environment, heritage and culture, a social context that is supportive of tourism (a history of working in the tourism industry and a welcoming attitude towards tourists) and the availability of tourism infrastructure and investment capital. Embeddedness serves to confer local distinctiveness which provides a competitive advantage or rent for local products (Hinrichs, 2000). Effective disembedding through promotion and marketing is required, however, to attract tourists from external markets because local demand is rarely adequate to sustain tourism in lagging regions (Cawley *et al.*, 2002). Tourism should also be 'complementary' to local economic and social structures if it is to contribute effectively to rural development (Pigram, 1980). Thus, rural tourism has been promoted throughout the EU as a method of compensating for decline in income from agriculture and fishing, following reform of the Common Agricultural Policy (CAP) and the Common Fisheries Policy (CFP). Tourism may be complementary also in supporting existing policies for the conservation of local resources and may establish an economic motive for such conservation. There is a possibility, however, of conflict arising between tourism and local cultures, if the integrity of the latter is threatened in any way by the former (Briassoulis, 2002). Part of the complementarity of tourism relates also to the scale at which it is pursued. Rural localities where agriculture is declining and few other employment opportunities exist can rarely compete with urban destinations or beach resorts on the basis of scale economies. Neither should they seek to do so, because their competitive advantage lies in responding to the growing demands for the alternative experiences that they offer (Butler *et al.*, 1998). Smallness of scale is a characteristic feature of tourism in rural locations and refers to the size and extent of tourism resources and the volume and impact of tourism activities vis-à-vis the existing resource base (Lane, 1994). Integrated tourism should be conducted at an 'appropriate' scale with regard to local structures (Ray, 1998a). This usually involves locally-owned small-scale accommodation properties and recreational pursuits engaged in individually or in small groups, which are most conducive to environmental conservation (Urry, 2002). Opportunities for contact with local people and local cultures frequently provide a strong motivation for visiting rural areas.

An integrated approach to tourism development in rural areas requires networking between the various actors involved. According to Murdoch (2000: 408) a network approach "can straddle diverse spaces and can hold the promise of a more complex appreciation of 'development' than has traditionally been evident in state-centred market-led or endogenous versus exogenous models". He was thinking in particular of the changing policy context in which food production takes place but tourism equally involves local and extra-local networking. Courlet and Pecqueur (1991) similarly identified the growing importance of concepts of partnership among stakeholders in thinking relating to deconcentration and regional development in France in the late 1980s. In the SPRITE project, networking was conceptualised in the context of local horizontal networks and extra-local vertical networks. Horizontal networking is central to the local creation of tourism businesses and vertical networking is necessary to source state support and attract tourists.

Integrated tourism is closely linked into a normative conceptualisation of sustainability. Being socially constructed, 'sustainability' lacks precise definition and is best seen as a process (Aronsson, 2000; McCool and Moisey, 2001). The details of sustainability are place-specific and subject to debate. However, a key objective of promoting tourism in lagging rural regions is to support the local economy and society and to do so in ways that protect rather than threaten the inherent characteristics of the local culture and the inherent quality of the natural environment. In this way, sustainability is multi-dimensional. Entrepreneurs and resource controllers are well-placed to gauge the extent to which these multi-dimensional objectives are being achieved. There is increasing acknowledgement also that rural development processes should empower local people through their involvement and through delivering benefits to individuals and the community more generally, thereby facilitating further development (Sofield, 2003).

Study Area, Data Collection and Analysis

Study Area

The reported research focused on the West Region of Ireland, in particular the coastal and mountainous areas of west County Mayo and northwest County Galway (Figure 8.2). The region's tourism is based closely on the natural environment and cultural resources which include parts of the Gaeltacht (an extensive area of west County Galway, and small areas on Achill Island and the Erris peninsula in the extreme northwest of County Mayo, where a majority of the population speak the Irish language). The region was selected for study because of its history of tourism, which dates to the end of the nineteenth century, its nationally and internationally recognized image as a destination, and its reputation as a repository of Irish culture. Four sub-areas were selected for study within the wider region, one in County Galway and three in County Mayo (Figure 8.2). Touring by car, coach and bicycle, and walking, angling and golfing are popular pursuits in all of the areas but there is also some specialisation, notably game angling in the Moy Valley, religious tourism in the south of Central Mayo, and Irish-language based tourism in Achill Island-Northwest Mayo. The findings of the empirical research are discussed here in the context of the four sub-areas together and inter-area differences are noted where pertinent.

Selection of Actors, Data Collection and Analysis

The actors surveyed were identified in a purposive way, with advice from a Consultation Panel of experts established in association with the project, from a comprehensive list of all relevant businesses and resource controllers operating in the four selected sub-study areas. The criteria for defining a business were that it was: primarily profit seeking (although state and cooperative enterprises were also included); either directly or indirectly involved in tourism; relatively small in scale

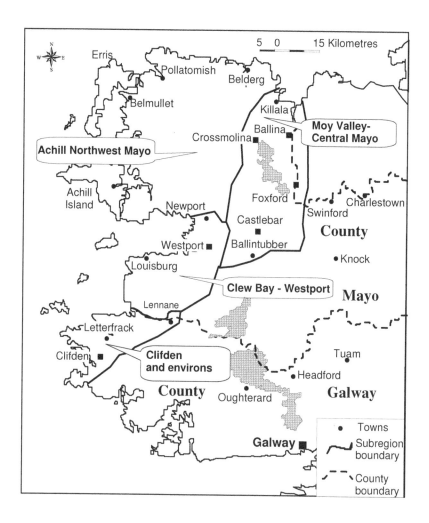

Figure 8.2   West Region of Ireland: sub-study areas © Ordnance Survey
Ireland/Government of Ireland. Copyright Permit
No. MP 006507

in terms of infrastructure; not oriented primarily at a mass market; and linked into the local physical environment, economy, society or culture. Large externally-owned hotels, large self-catering accommodation developments and coach tour companies were omitted because of not being conducive to the promotion of the inherent features of integrated tourism identified. However, their impacts, both positive and negative, were captured during the interviews with the other actors. The resource controllers were defined in terms of their ownership and (or) managerial control and (or) provision of resources and infrastructure for tourism. Efforts were taken to represent the different types of resources that were relevant to integrated tourism in the region: environmental (including aspects of the natural and built environment), economic, social and cultural. Cultural resources were defined as relating to the Irish language and the performance (music, song, dance, drama) and figurative (painting, sculpture) arts.

Fifty-one businesses were surveyed (Table 8.1). They consisted of accommodation and catering enterprises (small hotels, Irish Tourist Board-registered and non-registered bed and breakfast premises (B&Bs) and self-catering accommodation, a hostel, restaurants, public houses); providers of recreational activities (adventure centres, golf clubs, health centre, equestrian activities, horse drawn caravans, diving, angling, sailing, walking); heritage and cultural centres; and other businesses that included craft shops, a knitwear manufacturer and ferries. They included both well-established (eleven predated 1980) and newer enterprises. Representatives of twenty-three resource controllers were interviewed which included units of national organisations, autonomous regional bodies, local area partnerships, and one farm business (Table 8.2). The majority were in the state or semi-state sectors but a voluntary body, a private business, a cooperative and a public-private partnership were also selected. The main tourism planning, promotional and training agencies were included. Efforts were taken to represent local, regional and national dimensions of resource control, as pertinent. The interview procedure followed was primarily qualitative and the text was analysed in a thematic way (Feldman, 1995). Because of the focus on common key issues in both sets of interviews, cross-comparison was possible between responses for the two groups. This information is used here to: (i) profile the two groups according to the extent to which they reflect the selected dimensions of integration in their objectives and actions; and (ii) to analyse their respective views relating to the forms of tourism that are most appropriate to develop in the region and the ways in which these align with or depart from the concept of integrated tourism. In this way it is possible to arrive at an assessment of the potential offered in reality by the culture economy approach advocated in the SPRITE project as a method of contributing to rural sustainability and empowerment.

Table 8.1   Type and location of businesses (numbers)

Business type	Location				Total
	Clifden area	Westport area	Achill Island/ Northwest Mayo	Moy Valley/ Central Mayo	
Accommodation	7	0	8	5	20
Activity	5	2	6	3	16
Visitor Centre	2	3	0	2	7
Other	1	4	1	2	8
Total	15	9	15	12	51

Table 8.2   Organisational types interviewed

Organisations that control resources for tourism (n in brackets)
State training and employment agencies (3)
State resource and research agencies (6)
Regional Tourism Authority and Údaras na Gaeltachta (2)
Local Authorities (4)
Local Area Partnerships (6)
Tidy Towns Committee (1)
A farmer (1)

Characteristic Features of the Businesses and Resource Controllers

Both actor groups possessed a range of characteristics, in addition to the criteria on which they had been selected, that aligned closely with the features of integration as proposed in the conceptual framework. Local endogenous ownership of businesses was the norm, most businesses used local imagery in their promotion and marketing, and many reported having used personal and family resources in establishing their enterprise (although several had obtained external grant assistance from state agencies and the EU). The resource controllers were principally branches of exogenous state agencies but they were important conduits for national and EU investment in road and marine infrastructure, leisure facilities and interpretative centres, which enhanced the region's attraction for tourists. Both sets of actors were deeply embedded in the region; the businesses through their use of the physical environment for tourism purposes, and the resource controllers through their management role. Some of the resource controllers had become more closely involved in tourism during the latter half of the 1990s, when rural development was formally adopted as the 'second pillar' of the CAP (CEC, 1996b). Both actor groups, and particularly the businesses, sourced labour, materials and services locally. Most of the businesses engaged in successful disembedding to reach external markets and attract tourists and were assisted by particular resource controllers. The majority of the businesses operated on a small scale and the expansion that had been pursued was sympathetic to the environment in which they were located. The resource controllers were supportive of tourism activity of this type. Seasonality was identified as a weakness but the seasonal supplementary income was valued highly. Another aspect of complementarity identified related to the benefits gained by the local community from tourism infrastructure (for example, restaurants, leisure facilities, ferries). Active horizontal networking was the norm. Most businesses were members of at least one promotional and marketing organisation (national organisations, the county branding organisation, Mayo Naturally, and local groups), they liaised with the Regional Tourism Authority (RTA) in the conduct of promotion and marketing, and many referred clients to other local businesses. External vertical networking was engaged in extensively by both actor groups when sourcing funding and advice, in promotional and marketing activities and, by the resource controllers, in representing their own or members' interests.

Promoting the economic sustainability of their own enterprise was the main objective of most of the entrepreneurs and they also cited contributing to social and economic sustainability more generally. Many resource controllers had specific environmental management responsibilities and several highlighted a role in promoting local empowerment through remits relating to economic and social development and training. The ownership profile of the businesses indicated that empowerment through entrepreneurial opportunities was being provided for women in areas where few alternatives were available (24 of the 51 entrepreneurs were women).

Appropriate Forms of Future Tourism Development

Both actor groups were asked a number of open questions about the types of tourism development that they felt were most appropriate for the region in the future. Their respective views were in broad agreement and corresponded closely with the definition of integration being used in the SPRITE project. Offsetting deficiencies in integration was advocated in certain instances and reflected a concern with the more effective pursuit of a culture economy approach.

Local business ownership and natural and cultural resources were identified as providing potential for further development of integrated tourism by both sets of actors. The entrepreneurs highlighted the importance of promoting a quality personal service for business success. One fisherman's wife and B&B operator in the Clifden sub-area described this service as follows: "We give time to people. J is a good skipper and loves his work. We look after the guests." Several entrepreneurs expressed dissatisfaction with the perceived level of support for investment and promotion provided by the RTA, which was said to favour larger businesses. This perception is correct but the underlying explanation is imperfectly understood by the entrepreneurs; namely, that the role of supporting investments of less than €175,000 passed from the RTA to LEADER partnerships in the early 1990s. The entrepreneurs were more aware than the resource controllers that farmers were under-represented among tourism providers and of their role in constructing a 'rural' product. Both groups perceived opportunities to further develop leisure activities and experiences based on the natural environment and local culture and thereby increase embeddedness: "(Would) like to attract special-interest types. Would like more to come for archaeology, folklore, natural environment, architecture" (manager of a heritage centre, Westport area). The potential provided by the region's resource base for future tourism development, as perceived by most of the resource controllers, emphasized embeddedness in the natural landscape and was summarized by the manager of the Mayo National Park as follows: "Landscape, natural beauty, outdoor pursuits, angling and walking." Some entrepreneurs perceived that opportunities existed to use regional and local branding more effectively, as a method of capitalising on the existing regional image and reducing fragmentation in promotion and marketing. All respondents were aware of the need to continue to disembed to attract tourists to visit the region.

There was general support for small-scale quality tourism development both in terms of infrastructure and visitor numbers. The entrepreneurs were most aware that regional tourism was at a pivotal point where a choice existed to either focus on numbers or on more discerning tourists and most favoured the latter strategy. "It could develop into a high quality destination or it could go the route of cheap mass tourism. Large numbers would give a low yield and damage the environment" (hotel manager, Clifden). In this context, negative reference was frequently made to extensive tax-incentive self-catering developments located in Achill Island and Clifden, which were said to have captured business from existing accommodation providers, to be of poor design and to be 'ghost towns' in winter. In the words of one local B&B owner, "The locals do not own them. Achill Island should not have little housing estates, they do not belong." Some resource controllers with environmental conservation remits advocated stricter control of physical developments, citing incidents of hillside erosion and watercourse pollution. In general, however, tourism was viewed as being complementary to, rather than in conflict with, local social and economic structures. "It is the livelihood of the majority, particularly as farming is declining" (local authority engineer, Clifden). Intervention to promote complementarity was advocated by many respondents to increase access to the countryside. A 'right to roam' does not exist in Ireland and farmers are becoming increasingly concerned about liability for accidents occurring on their land. (A national Council for the Countryside was established in 2004 to address this contentious issue but a solution acceptable to all interests had not emerged by autumn 2006.) Many entrepreneurs identified issues where the local planning resource controllers were not catering adequately for the needs of tourism. Deficiencies in physical infrastructure (roads, water and waste treatment) were said to inhibit certain types of development and, with poor road signage, to militate against the wider circulation of tourists into the more peripheral areas. The resource controllers were more vocal in recommending stricter planning control to reduce the numbers of tour buses on scenic routes and peak season traffic congestion in Clifden and Westport, which impacted negatively on the quality of life for local residents.

Improved networking among local businesses was identified by both actor groups as being necessary in Northwest Mayo in order to develop a cohesive tourism product and promote the area effectively. "They work on their own now, but they need to look at other businesses as complementary" (owner of a self-catering unit, Achill Island). A number of state agency representatives with responsibilities for tourism training underlined the need for the collaborative development of tourism products in more remote areas, as expressed by one: "Local groups will need to develop types of packages; for example, links on a chain along which tourism will move." Entrepreneurs tended to be more aware of deficiencies in networking in the context of promotion and marketing. In the Clifden sub-study area larger accommodation businesses were said to be reluctant to collaborate in promotional efforts with smaller businesses. Intervention by Chambers of Commerce and the newer LEADER partnerships was advised, in order to promote local networking more effectively.

Potential Outcomes for Multi-dimensional Sustainability and Empowerment

The entrepreneurs almost universally perceived a need to promote quality characteristics as a method of developing distinctive tourism products that could command a price premium and contribute to growth in employment and income. They viewed sustainability largely in economic and social terms, although they were aware of the importance of protecting the quality of the physical environment in which tourism takes place. The resource controllers were generally supportive of integrated tourism and were concerned about protecting the quality of the physical (and to a lesser extent the built) environment as a public good which would require stricter control of certain types of tourism development. Both sets of actors, but particularly the entrepreneurs, were conscious of the importance of involving farmers more fully in local tourism developments, especially because of their role as stakeholders of a resource to which access is sought for hiking. The resource controllers were most conscious of a number of impending threats to the promotion of integrated tourism which highlighted a need for stricter control of development: a proposed gas terminal in Northwest Mayo, fish-farming in Clew Bay and, more generally, 'one-off' rural housing (individual dwellings on discrete sites). The entrepreneurs were opposed to further large-scale self-catering accommodation developments because of fears of capture of business. Some entrepreneurs in Northwest Mayo expressed concerns about the growing use of English in encounters with tourists and referred to a need to further promote Irish language-based tourism.

Both actor groups envisaged contributions to the empowerment of local people accruing from the types of tourism proposed; through more employment being generated locally, further provision of skills and training, and increased local expenditure by tourists. The Mayo National Park representative referred to the establishment of a local liaison committee to involve local people in decision-making relating to the designation of landscapes for conservation.

Conclusions

Tourism has been identified in recent EU and national rural development policies as providing opportunities to compensate for declining incomes from agriculture and fishing in lagging rural regions. Contemporaneously, some of the mountainous and coastal regions that have been affected most negatively by CAP and CFP reform are exercising an attraction for tourists who seek their innate natural and cultural attributes as an alternative to the everyday pressures of urban living and mass tourism experiences. The most appropriate forms of tourism development for such areas and the most effective methods of support and promotion remain issues for discussion. The SPRITE project addressed these issues by developing the construct of integrated tourism, drawing on concepts of culture economy, and by investigating the potential to pursue this approach in particular research contexts. Integrated tourism was defined as involving seven dimensions which were explored through the research. This chapter focused on the experience in the West Region of Ireland, which has a long history of tourism based on landscape and culture. The analysis reported referred to the two actor groups most closely involved in the production of tourism products and experiences, tourism businesses and resource controllers.

The results indicate that tourism in the West Region currently, and in terms of the preferences of both businesses and relevant resource controllers, broadly follows a culture economy pathway as discussed in recent literature. Selected local markers have been valorised for tourism purposes over time. These markers are encapsulated within the criteria defined in the SPRITE project: namely, endogenous ownership; embeddedness in the local environment, economy, culture and society; complementarity to more traditional sources of income; and small-scale development. Developments to date and advocated for the future are generally supportive of multi-dimensional sustainability and are conducive to the empowerment of local people as the evidence illustrates. The promotion of integrated tourism and the capture of cultural resources for economic purposes is not entirely unproblematic. Thus, the business owners tended to prioritize economic sustainability whilst the resource controllers were more conscious of a need for environmental sustainability, a divergence of views that requires attention if conflict is to be avoided over resource use. Extensive networking took place within and between both groups horizontally and vertically. However, a reluctance to network was identified among small-scale businesses in Northwest Mayo and between larger and smaller businesses in the Clifden area of County Galway, which require intervention if integration in development and promotion is to be pursued more effectively. In general it appears that local people are being empowered, particularly through the employment gained and the skills acquired. Nevertheless, some owners of smaller business felt relatively neglected by the RTA, by comparison with their larger-scale counterparts, arising from changes in support policies. Farmers were identified as experiencing disempowerment because of increasing pressure to provide access to their land for recreational purposes, without receiving what they consider to be adequate compensation or indemnity against legal action for accidents incurred. There are therefore several issues that require attention by local actors and policy-makers if integration is to be pursued more effectively.

A culture economy approach is in place in the study region but it appears that this approach has not yet been capitalized on fully, particularly with regard to external promotion and marketing. Regional markers have been valorised and the region has an established image as a tourism destination. Some businesses and resource controllers are involved in membership promotional groups and some are members of national organisations that award quality marks. However, there was limited evidence of the development of quality marks associated with the regional or local territory and therefore limited cohesion in external promotion and marketing. The Mayo Naturally branding organisation is active in promoting the county as a tourism destination but strict quality criteria have yet to be developed. The further development of regional branding was recommended by several of the entrepreneurs. If associated with quality attributions, such branding would serve to protect the inherent quality of the resources on which tourism depends, would contribute to more effective promotion and marketing and would promote longer-term economic and socio-cultural sustainability. It seems that this is the next step that should be taken in promoting integrated tourism in the region which would follow the second mode of culture economy proposed by Ray (1998a).

Acknowledgement

SPRITE was a collaborative research project conducted under the EC Fair 5 Programme. There were six international partners in the project based at universities and research institutes in Ireland, the UK, the Czech Republic, France, Greece and Spain. Dr. Tim Jenkins, Institute of Rural Studies, University of Aberystwyth was Co-ordinator. Róisín Kelly's assistance with the research at NUI, Galway is acknowledged gratefully.

References

Aronsson, L. (2000) *The Development of Sustainable Tourism*. London: Continuum.

Barnes, T. (2005) Culture:Economy. In: Cloke, P.J. and Johnston, R.J. (eds) *Spaces of Geographical Thought*. London: Sage, pp. 61–80.

Bossevain, J. (1996) *Coping with Tourists: European reactions to mass tourism*. Oxford: Berghahn Books.

Briassoulis, H. (2002) Sustainable tourism and the question of the commons. *Annals of Tourism Research*, 29(4): 1065–85.

Butler, R., Hall, C.M. and Jenkins, J. (1998) *Tourism and Recreation in Rural Areas*. Chichester: Wiley.

Cawley, M., Gaffey, S. and Gillmor, D.A. (2002) Localization and global reach in rural tourism: Irish evidence. *Tourist Studies*, 2(1): 63–86.

CEC (Commission of the European Communities) (1996a) *CdR 54/96 Opinion of the Committee of the Regions of 18 September 1996 on promoting and protecting local products – a trump card for the regions*. Brussels: CEC.

CEC (1996b) *The Cork Declaration: a living countryside*. Brussels: CEC.

CEC (2001) *Environment 2010: Our Future, Our Choice*. Brussels: CEC.

Courlet, C. and Pecqueur, B. (1991) Local industrial systems and externalities: an essay in typology. *Entrepreneurship and Regional Development*, 3(4): 305–315.

EUROPA (2003) *Press Release, Conclusions of the Second European Conference on Rural Development in Salzburg*, http://europa.eu.int/comm/agriculture/events/ Salzburg.

Feldman, M. (1995) *Strategies for Interpreting Qualitative Data*. London: Sage.

Government of Ireland (1997) *Sustainable Development, A Strategy for Ireland*. Dublin: Stationery Office.

Hinrichs, C. (2000) Embeddedness and local food systems: notes on two types of direct agricultural market. *Journal of Rural Studies*, 16(3): 295–303.

Ilbery, B.W. and Kneafsey, M. (1998) Product and place: promoting quality products and services in the lagging rural regions of the European Union. *European Urban and Regional Studies*, 5(4): 329–41.

Jenkins, T. (2000) Putting postmodernity into practice: endogenous development and the role of traditional cultures in the rural development of marginal regions. *Ecological Economics*, 34(3): 301–14.

Jenkins, T. and Oliver, T. (2001) *SPRITE Deliverable 1. IT: A Conceptual Framework*. Aberystwyth: Institute of Rural Studies, University of Wales, Aberystwyth.

Kemm, M.S. and Martin-Quirós, M.A. (1996) Changing the balance of power: tour operators and tourism supplies in the Spanish tourism industry. In: Harrison, L.C. and Husbands, W. (eds), *Practising Responsible Tourism*. Chichester, Wiley, pp. 126–44.

Kneafsey, M. (2001) Rural cultural economy: tourism and social relations. *Annals of Tourism Research*, 28(3): 762–83.

Lane, B. (1994) What is rural tourism? In: Bramwell, B. and Lane, B.V. (eds), *Rural Tourism and Sustainable Development*. Clevedon, Channel View Press, pp. 7–21.

Lash, S. and Urry, J. (1994) *Economies of Signs and Space*. London: Sage.

McCool, S.F. and Moisey, R.N. (eds) (2001) *Tourism, Recreation and Sustainability*. Wallingford: CABI.

Murdoch, J. (2000) Networks: a new paradigm of rural development. *Journal of Rural Studies*, 16(4): 407–19.

Organisation for Economic Co-operation and Development (1995) *Niche Markets as a Rural Development Strategy*. Paris: OECD.

Pigram, J. (1980) Environmental implications of tourism development. *Annals of Tourism Research*, 7(4): 554–83.

Ray, C. (1998a) Culture, intellectual property and territorial rural development. *Sociologia Ruralis*, 38(1): 3–19.

Ray, C. (1998b) Territory, structures and interpretation- two case studies of the European Union's LEADER I Programme. *Journal of Rural Studies*, 14(1): 79–87.

Ray, C. (1999) Towards a meta-framework of endogenous development: repertoires, paths, democracy and rights. *Sociologia Ruralis*, 39(4): 521–37.

Sage, C. (2003) Social embeddedness and relations of regard: alternative 'good food' networks in south-west Ireland. *Journal of Rural Studies*, 19(1): 47–60.

Shaw, G. and Williams, A.W. (2004) *Tourism and Tourism Spaces*. London: Sage.

Skuras, D., Meccheri, N., Moreira, M.B., Rosell, J. and Strathopoulou, S. (2005) Entrepreneurial human capital accumulation and the growth of rural businesses: a four-country survey in mountainous and lagging areas of the European Union. *Journal of Rural Studies*, 21(1): 67–79.

Sofield, T. (2003) *Empowerment for Sustainable Tourism Development*. London: Pergamon.

Terluin I.J. and Post, J.H. (eds) (2000) *Employment Dynamics in Rural Europe*. Wallingford: CABI.

Urry, J. (2002) *The Tourist Gaze*. London: Sage.

World Commission on Environment and Development (1987) *Our Common Future*. Oxford, Oxford University Press.

Chapter 9

The Cumbria Hill Sheep Initiative: A Solution to the Decline in Upland Hill Farming Community in England?

Lois Mansfield

Introduction

Upland and mountain areas operate on the fringes of viable agricultural production. Through their physical constraints of soil, climate and topography these environments limit English farmers to livestock production with typical profit margins of around £5000 per annum, well below the national United Kingdom (UK) average (Chadwick, 2003). Compounding these low economic returns are problems of succession amongst the farming families and an increasing mismatch between production and the post-productivist vision for British agriculture (MAFF, 2000).

The uplands of Cumbria in northern England are no exception to these issues. A system of farming has developed here to make the best use of the environment by adapting farming practices to fit the harsh climate (2000mm per annum and a growing season of temperatures above 5.6°C for less than 190 days; Grigg, 1995) and rugged terrain. The landscape produced by this activity has become highly valued in terms of semi-natural ecological communities and recreational pursuits (Cumbria County Council, 1997). Indeed, it forms a core feature of the Lake District National Park and the Park Authority's bid to secure World Heritage status (Chitty, 2002). The problem is that the farming system itself may not survive to sustain this desired upland landscape. This chapter explores a package of measures, known as the Cumbria Hill Sheep Initiative (CHSI), introduced by the local Rural Community Council to aid in the maintenance of upland agriculture.

The Character of Upland Agriculture in Cumbria and Associated Issues

To understand why the CHSI was initiated it is important to understand something about the character of the farming system and the nature of the problems it faces in Cumbria.

The Upland Farming System

The farming system of the Cumbrian uplands is a product of many thousands of years, culminating in a landscape not unlike that shown in Figure 9.1. This farm landscape comprises three distinct land types: inbye, intake and fell.

Inbye land is by far the best land, close to the farm buildings and used for the production of hay or silage for the winter, grazing land in winter months and lambing areas in spring. At the other extreme are the *fells* at the highest altitudes (usually 300m ASL or more). These are areas typically of heather moorland or rough unimproved grass pasture highly prized in terms of nature conservation in the UK and Europe (English Nature, 2001; Thompson *et al*, 1995). Indeed, it is the agricultural management of the land in the past that has allowed these ecological communities to develop through extensive grazing regimes and periodic burning of the heather (*Calluna vulgaris*) to re-invigorate growth (Backshall *et al.*, 2001). In between the fells and the inbye lies the *intake*, sometimes referred to as *allotment*. This is land that has been literally taken in from the fell and enclosed commonly using drystone walls made of locally field cleared stone.[1] The last wall before the open fell is known as the '*fell wall*'. The system of walls, enclosed fields and fell areas are then what give the uplands of Cumbria their intrinsic high quality landscape so desired by the public (Cumbria County Council, 1997; Ratcliffe, 2002).

Farmers run mainly two enterprises in the core of the Cumbrian uplands – sheep and/or beef; on the valley bottoms and upland margins some environments are sheltered enough to run a dairy herd. Occasionally farms even run a dairy herd and a fell sheep flock, although this is labour intensive. Upland farms, themselves, are divided into two types; true *upland farms* containing inbye, intake and fell and the *hill farm*, which contains intake and fell with little or no inbye. This tends to restrict hill farms to traditionally running just sheep, whereas the true upland farms have historically run sheep flocks and cattle herds in combination.

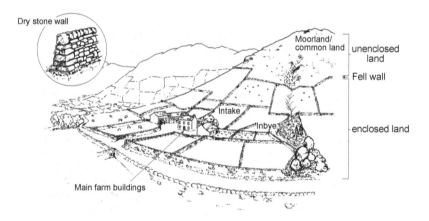

Figure 9.1   A typical Cumbrian hill farm landscape

From the farmers' point of view the landscape they have developed has a number of functions. Walls keep livestock from straying; they keep rams away from ewes at the wrong time of year and they allow stock to be grazed in winter on a rotational basis to ensure sustainable grassland management. The fell areas are summer pasturage, when the enclosed land's productivity has been exhausted or allocated for the production of grass and hay crops for winter feed. In order to support the same number of sheep on the fell as in the inbye, the lower productive land needs a substantially larger area over which the sheep disperse. This grazing area has developed over many generations of farmers, who originally shepherded the sheep, keeping them to land over which the farm had common rights. Over time the sheep get to know the land that they can graze on and gradually the intensive shepherding can be withdrawn so that the flock manage themselves geographically. This instinct of the sheep to keep to a certain land area is known as '*hefting*' or '*heafing*', the operation of which can vary from upland to upland (Hart, 2004). The ewes pass the knowledge of the area (heft) on to their lambs, who in turn pass it on in turn to their lambs. In this way it is important that the farmer maintains a multi-generational flock. Typically a common in Cumbria can be many thousands of hectares of land and thus can contain enumerable hefts (Figure 9.2) isolated from the main farm unit. Over time the *virtual* boundaries between hefts have developed keeping stock from straying into another heft, thus developing a self policing of grazing pressure.

Another important feature of the upland farm system is the '*gather*'. Sheep are collected and gathered together from the open fell at various times of year and brought down to the farm for shearing, worming, winter grazing, sales and lambing. (Few farmers lamb their sheep out on the fell now for management reasons). Because hefts are geographically extensive, over difficult terrain, the labour requirements for gathering are high (as many as 25 people for a single gather). This is exacerbated by precipitous landscapes that do not lend themselves to modern All-Terrain Vehicles, thus pedestrian access is often the only means reaching the spread out stock:

> These fells have been shepherded. They're shepherded the way now as they were 200 years ago with a dog and a stick. You know, there's no flying around on motorbikes or whatever on the high fells so they've got to be managed as they were years ago. (Farmer 5, Burton *et al.*, 2005)[2]

Traditionally, farmers, their families, staff and sheep dogs work together over an entire common (several hefts) to gather several flocks in one day. In this way a large number of people work cooperatively to clear all sheep from the common in an efficient manner (Burton *et al.*, 2005). Upland commons in Cumbria can be extensive, the common shown in Figure 9.2 is around 8,850 hectares with the heft indicated being about 150ha (Aitchison *et al.*, 2000) and thus cooperation between people is essential if all sheep are to be brought down safely. The sheep are then divided into the distinctly owned flocks down at the fell wall either there and then, or through the '*Shepherds Meet*', a separate event when mis-gathered sheep are exchanged between farmers. Given the labour intensive nature of the gather, any losses in farm labour are difficult to manage as short-term contractors have neither an intimate local knowledge of each unique fell nor an understanding of the behaviour

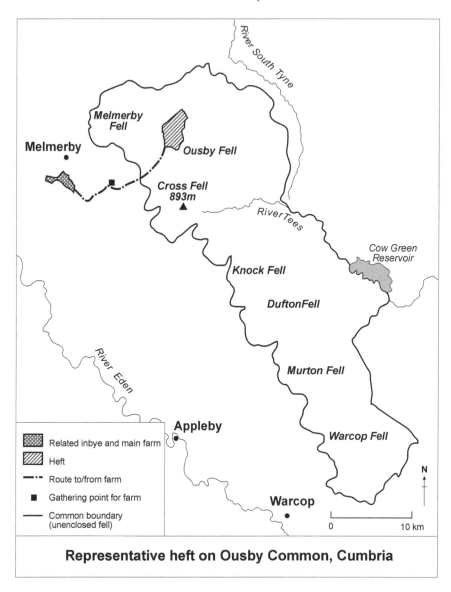

Figure 9.2   Representative heft on Ousby Common, Cumbria

of the sheep, and they lack familiarity with the land itself. Furthermore, experience accrued over years also enables the farmer to recognise where sheep will be in times of adverse weather conditions.

The Issues

Whilst this farming system has emerged to take effective use of the upland environment and the resources it can offer, changes in economic and political circumstances have increasingly marginalised hill farming (Haskins, 2001). As profits have declined farmers have had to make some tough decisions as to how they can continue to operate. Initially, farmers have three options:

1. To tighten one's belt and continue with ever decreasing profits;
2. They can diversify, taking up the variety of grants and incentives on offer to them;
3. They can withdraw from farming altogether.

However, each of these three options does create further problems.

If the farmer chooses to continue to farm in a similar way, he/she must seek mechanisms to reduce costs. Typically the easiest way to do this has been to reduce the paid labour force on the farm. Many farms now rely solely on the farmer and the partner for labour, with older children helping out when they can. For some hill farmers, they cannot cut the wage bill as they are not married, do not have children or their partner already works off-farm. For example, with the farmers involved in one of the schemes described below, the Fell Farming Traineeship Scheme, 80 per cent have no partner or children to directly support them with the day to day running of the farm.

Whilst cutting labour saves money in the short term, in the long run it can cause problems for certain aspects of the farm management. One particular issue is the lack of people at gathering times to control the behaviour of flocks as they come off the fell. One farm on Ousby Fell used to have 22 people going out for the gather; they are now down to nine (Burton *et al.*, 2005). From a practical management point of view there are simply not enough people to close off the escape routes for the sheep leading away from the main flock. Another problem of this lack of labour is that it limits strategic investment in new non-farm enterprises.

The second option is for the farmer to diversify their enterprise base. Some farm families have successfully developed bed & breakfast enterprises or bunkhouse barns. One family in the Lake District had 50 people going through their bunkhouse over the two-week Christmas period in 2004 (Mansfield and Martin, 2004). Perhaps one of the easier ways of diversifying has come from the adoption of agri-environment initiatives such as the Environmentally Sensitive Areas (ESA) or Countryside Stewardship Schemes (CSS). All farmers involved in the CHSI had taken up one or other of these schemes. However, whilst it can be argued that adoption of ESA or CSS involved little capital cost to hill farmers due to their starting environment, many do not have enough disposable income to make the leap to more complex and demanding types of management which command higher payments, or simply choose not to invest. Financial constraints also limit other diversification opportunities through matched funding requirements or lack of surplus labour.

The third option open to the hill farmer is to escape the poverty trap of the cost-price squeeze and opt to withdraw from farming altogether. A number of farmers have done this, spurred on by the effects of Foot and Mouth disease in 2001 (Franks *et al.*, 2003). Some have sold up altogether, others sold off the land only. Either situation has multiplier effects into the external farm environment of the landscape and the wider community. Those that have sold up altogether have often split the house from the land. This has happened in south Cumbria around the Howgills where 45 per cent (17 out of 36) of the farm units are no longer farming (personal communication, H. Wilson). The effect is two-fold: first, the household becomes disenfranchised from the farming community and, second, the land can be abandoned. If the latter happens on the heft the associated de-stocking affects surrounding hefts, whose sheep move into the new unclaimed territory. On large fells like the one shown in Figure 9.2, the ripple effect of heft abandonment can affect tens of farms and their management of the flocks, particularly at gathering times. The abandonment of the hefts also leads to problems for the semi-natural vegetation. Because pressure for grazing has been lessened, sheep can graze more selectively, eating out the sweeter and more nutritious species at the expense of the less desirable. Plants such as gorse (*Ulex europaea*), bracken (*Pteridium agustifolium*) and mat grass (*Nardus stricta*) have increased, leading to a deterioration in the semi-natural vegetation on open fells when not managed (Backshall, 1999).

Underlying all these issues of the re-structuring of the farm management and its enterprises is the subject of labour. The cost-price squeeze has reduced the numbers of the agricultural workforce on upland farms, limiting the farmers' opportunities to invest in diversification, their ability to increase profits or in some cases to actually secure succession. The agricultural workforce has been contracting for many years, a trend exacerbated by the fact that the farming population is getting older, with fewer young people coming into the industry (DEFRA, 2004). Furthermore, succession is a fundamental part of the hill farming system as it allows for intergenerational skills to be passed on (Gray, 1998).

This then, is the situation facing upland farming in Cumbria; a cost-price squeeze with few options allowing escape and a lack of succession, which threaten to lead to the collapse of the system. It was these issues and the effects of the aftermath of the Foot and Mouth epidemic which led to the design of the CHSI.

The Cumbria Hill Sheep Initiative (CHSI)

The CHSI was devised by Voluntary Action Cumbria[3] in conjunction with the main land use organisations and the farming community in Cumbria. Its goals are to re-invigorate the economic potential of the hill farming industry and re-align its priorities in line with current rural policy coming from the EU. Funding has been provided from a variety of sources including the National Trust, the Lake District National Park, English Nature and DEFRA. However, the most substantial component of funding has come from Cumbria Fells and Dales LEADER+ (*Liaison Entre Actions de Development de l'Economie Rurale*), fulfilling its remit to add value to local products under Theme 3 of EU LEADER+ and Objectives 1 and 2 of UK LEADER+[4] (LEADER+, 2000). The LEADER approach of encouraging bottom-up rural development, rather than top-down strategies, is of particular value to the Cumbrian situation given the conservative character of the farmers and the geographically diverse nature of the problems.

The CHSI is an umbrella project comprising a number of separate schemes that overlap to accommodate the issues facing the hill farming industry as a whole. Two of the projects are county wide across Cumbria:

1. *The Fell Farming Traineeship scheme* – designed to tackle the succession issue;
2. *The Social Capital in Hill farming research project* – developed to investigate the non-economic value of hill farming.

The third project focuses on the geographically indigenous Cumbrian sheep breeds. Each of the three Breed groups (Rough Fell, Herdwick and Swaledale) has developed a range of initiatives designed to raise the profile of the breed and to seek out new marketing opportunities.

The Fell Farming Traineeship Scheme

The Fell Farming Traineeship Scheme (FFTS) was developed in 2002 through the collaborative efforts of a range of land management organisations[5] in Cumbria. Their overall aim was: 'To tackle the issue of providing the effective transmission of the culture (skills, knowledge and understanding) of hill farming in a situation where the labour market is falling' (Mansfield and Martin, 2004: 5). In other words, to address the dwindling labour force in upland agriculture, by supporting young people through the provision of skills training and a wage. The project centred on the training of young people between the ages of 16 and 30, who would otherwise have restricted opportunities to enter into the hill farming sector and who would be lost to other occupations. The hope was that these young people would remain in the sector and take over farms as the older generation retired.

Each trainee was placed with a cluster of farmers (Figure 9.3). Each 'farmer ring' had a lead farmer who mediated the patterns of work between the farms in the ring. The farm ring comprised of four or five farms, each of which had a different enterprise mix. For example, one farmer ring in the Bampton area of the Lake District comprised a cattle fattening enterprise, a farm with good environmental diversification, a Belgian Blue embryo transplant operation and the fourth had a wide range of agricultural machinery. All farms had a hill sheep enterprise as well. The trainee then moved around the farms to an agreed pattern of attendance. During their time on the farm, each trainee worked with the farmer and any other staff to learn stock tasks, estate and machinery skills in context. At various points the trainee was taken off the farm and put through a range of competence-based certificated courses or training days such as sheep dipping, machinery certification, fencing and walling, and agri-environmental scheme briefings.

Trainees worked from 9 to 5, completing a 37-hour week in line with British employment law. However, this became problematic as it did not truly reflect the farmer's working week:

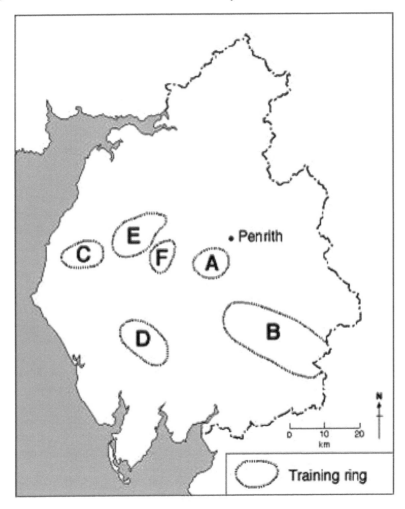

Figure 9.3   Geographical distribution of trainees and farmer rings

> You can't leave off fencing it at 5pm to let the stock wander out into the road, nor go
> home at five and ask the ewes to cross their legs and not lamb until 9am the next morning.
> (Farmer 2, Ring A)

The scheme was monitored and evaluated throughout its operation with a view to
mainstreaming the initiative in other uplands farming areas in the UK. A variety
of techniques were used including: training needs analysis (TNA); practical
competence assessment using National Vocational Qualifications criteria; an
environmental goods exercise, where trainees mapped extant conservation features
and then attempted to suggest improvements; informal farmer interviews (6) and
formal farmer questionnaire surveys (24). Trainee development and the succession
issue are discussed below.

Trainee Development

The most important aspect of each trainee's development was to evaluate whether they had acquired knowledge, understanding and practical hill farming skills. The initial TNAs showed that four of the trainees had little or no skill in the areas of estate skills and farm administration. All trainees had some skill in stock tasks. This pattern of skills is understandable as young people tend to develop, initially, skills in areas of livestock, as, according to the observations of the assessment team, they can assist in various stock tasks from an earlier age. In contrast, mechanical opportunities are somewhat restricted due to both age limitations and personal preferences of youngsters within a mainly livestock county.

The follow-up assessments were favourable but did identify some gaps. In all areas of skills assessment the trainees were competent on the practical side. However, theoretical understanding as why things were done in a certain way or how to choose appropriate construction materials was lacking. Machinery assessments demonstrated that the tractor was no longer seen as the primary machine on the hill farm, being superseded by the quad bike.

The greatest disadvantage of the scheme was the age of the trainee, where 27 per cent of farmers interviewed felt that they were too young. This led to problems of transportation for non-drivers in the more expansive clusters, limited ability due to age and physique, and too much being expected of young people by the funding bodies and stakeholders. There were also some inevitable but occasional problems with lateness, unreliability and inflexibility.

In terms of the environmental goods exercise, trainee knowledge was extremely variable. Table 9.1 summarises the outcomes of the environmental goods exercise. The main conclusion was that the two trainees who have had formal education in a land based area had much more of an idea than those who had not. The younger trainees failed to identify many extant habitats and had little idea how to enhance what they did identify. Many of the improvements suggested either had an agricultural bias or were conservative small scale activities, which were focused away from the agricultural land, such as planting up field corners. This led to developments which were isolationist, rather than integrated. Such separation is of concern as it is this integration that is central to the new Environmental Stewardship Scheme under the European Rural Development Programme (DEFRA, 2005). However, the farmers looking after the trainees were not surprised by this poor knowledge, due to the trainees' lack of experience, and they felt that some of the agencies who were supporting the scheme had too high expectations of the trainees from an environmental point of view.

The Labour Issue

The interviews and surveys of the 24 farmers identified three main areas of concern about labour which they hoped the scheme would address throughout its operation and beyond: succession, tenancy issues and farm management. Lack of succession was recognised by all the farmers as the biggest problem facing the industry. A high number of the farms in the scheme did not have an identified successor (80 per cent);

some as a result of tenancy arrangements, others because their children were not interested, that they were single or had no offspring. Thus, new people coming into hill farming through the FFTS were seen as essential by the scheme farmers.

Figure 9.4 shows that the greatest benefit identified by the farmers was for immediate labour relief (87 per cent). Longer term, the opportunity for skills and knowledge transfer was deemed important (39 per cent), followed by the fact that it helped to keep young people in farming and that it helped during 'gathering' (28 per cent each). In a broader context, about 50 per cent of the farmers explicitly said that hill farming relied on individuals to understand how to work with the *unique* environment of each farm. This observation has been corroborated through the Social Capital project (see below).

Tenancy was the second main concern as a number of the hill farms in the survey area are properties of the National Trust[6] (25 per cent), with a mixture of old and new tenancy arrangements, where the new tenancies do not allow for inter-generational inheritance. Examples were given by tenants that inappropriate people had been given tenancies in the last few years, who simply did not understand the need for cooperation in terms of stock management on the high fells. This had led to a number of incidents where sheep had been ineffectively gathered and dipping regimes ignored.

Table 9.1   Environmental goods exercise

	Trainee 1	Trainee 2	Trainee 3*	Trainee 4	Trainee 5	Trainee 6
Demonstrated an understanding of the basic principles of environmental management, e.g. low nutrient inputs				x	x	
Drew on linear features	x (some)	just streams		x	x	x (some)
Drew on areal features	x	x		x	x	x (some)
Recognised all current habitats		x		x	x	only open fell
Added new linear features				x	x	x
Added new areal features	x (very cautious)			x	x	
Identified agri-environment grant opportunities				x		
Could give detailed information about agri-environment initiatives				x		
Integrated agricultural and environmental practices across the farm				x		
Trainee profile (male/female) + (mature/young)	M + M	M + Y	M + Y	F + M	F + M	M + M

*   no response.

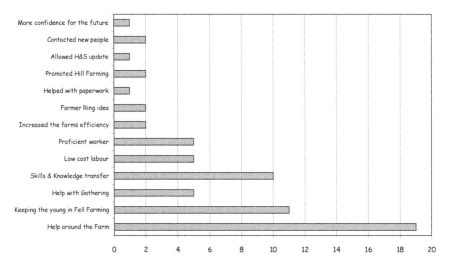

Figure 9.4   Benefits of the FFTS as perceived by participating farmers

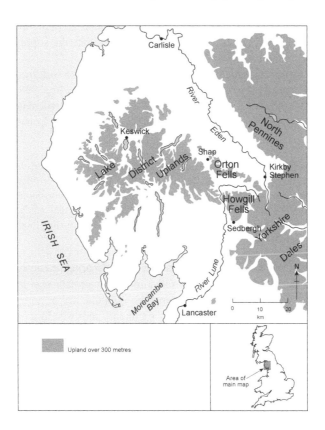

Figure 9.5   The Cumbrian Uplands

With respect to farm operations the largest problem perceived was related to gather management with the loss of labour. A particular problem was farm amalgamation, leaving a number of hefts operated by one family. Over half of the farms involved in the scheme now relied on only the farmer and their partner to complete this work. In the past all had employed labourers and shepherds full-time, but now simply could not afford to do so due to the low profit margins. The FFTS has provided the financial support for the majority (79 per cent) to benefit from an extra pair of hands. The reflections at the end of the programme reiterated this, where nearly half of the benefits identified were labour related (Figure 9.4). In terms of the FFTS, 60 per cent of farmers had commented that they had liked the FFTS, because it has shown them how useful extra labour was and that they can manage to pay for one day per week, but not a full-time position. Working with other farmers meant that, as a collective, they could employ someone full-time, which otherwise they could not have done individually. This has now happened with cluster B in the south-east of Cumbria (Figure 9.3).

Social Capital in Hill Farming

The second scheme forming the CHSI was the Social Capital in Hill Farming research project. This research was designed to investigate the supply and demand of the hill farming industry for public goods.

As with all forms of agriculture in the EU, there has been a shift from productivist to post-productivist activity over recent years (Ilbery and Bowler, 1998), culminating in EU Rural Development Regulation (RDR) 1259/99 and articulated in the UK through the four provincial Rural Development Programmes. Lowe *et al.* (2002: 15–16) note:

> The social justification both of modulation and of the various measures under Article 33 of the RDR is not so much agricultural survival as the provision of broader environmental public goods for a society that places particular value upon them. Similarly, farming's long-term role is that of developing and responding to particular market opportunities resulting from shifting social demands on the countryside (quality food, regional food chains, farm tourism and countryside management).

This has led to an emphasis on the production of non-market rather than market-led outputs (Table 9.2). Therefore the overall aim of the project was to demonstrate that the production of public goods requires the maintenance of an upland farming system, its communities and workforce. A secondary aim was to try to understand what the public's perception was of this relationship between agricultural activity and public goods.

Investigating the Demand Side

The demand side of the Social Capital research project focused on a cost-benefit analysis of the public's preferences of public goods in upland areas (McVittie *et al.*, 2005). Two focus groups (Skirwith, a village in Cumbria, and Sheffield) and 1000

Table 9.2   Market and non-market outputs from agriculture (adapted from: McVittie *et al.*, 2005)

Market Outputs	Non-Market Outputs	
Traditional Economy	**Environmental**	**Health & Safety**
• Food quality • Food quantity • 'Cheap' food • Employment • Farm Incomes • Rural multiplier • Export Income • Rural tourism	• Landscape • Habitats • Biodiversity • Water quality • Soil Conservation	• Animal Health & welfare • Prevention of animal to human infection • Food Safety • Managed genetic modification • Biosecurity
New Economy	**Cultural/traditional**	**Social**
• Recreation/access • Organics & other niches • Controlled appellation & regional origin	• Landscape • Traditional industry • Rural Character • Local food • Recreation • Prevention of urban sprawl	• Local employment and economy • 'Vibrant' communities • Tourism • Social cohesion • Educational resource

postal questionnaires (diverse areas of Cumbria and Manchester), were employed to explore the public's preferences regarding non-market goods in the Cumbrian uplands using Contingent Valuation Method (CVM), in this case Willingness-to-Pay, and Analytical Hierarchy Process (AHP) techniques (McVittie *et al.*, 2005).

McVittie *et al.* (2005) summarised the upland landscape in terms of three main attributes, to each of which they applied a number of qualities:

1. *upland landscape* – scenic views, traditional buildings, peace and tranquillity, wildlife;
2. *traditional farm management* – family farms, farming skills;
3. *community culture* – local culture and social networks.

The qualities and attributes, in pairs, were then graded by respondents on a sliding scale against each other as shown in Figure 9.6. The attributes and (the three qualities separately) were weighted according to their relative importance for the five groups of people (rural, remote rural, urban, Manchester and Cumbria) totalling up to one. The weightings were then statistically tested using means to see if significant differences occurred between the five sample groups. Results from this statistical analysis are shown in Table 9.3.

The AHP results shown in Table 9.3 demonstrate that the Manchester sample had a significant preference for the 'upland landscape' quality, in contrast to their Cumbrian counterparts who preferred 'traditional farm management' and 'community culture'. McVittie *et al.* (2005: 25) suggested that:

Table 9.3   Statistical test results for AHP attributes and qualities

	Cumbria vs. Manchester	Urban vs. Rural vs. Remote rural
Attributes		
Upland landscapes	\ * with Manchester	\
Farm management	\ * with Cumbria	\
Community culture	\ * with Cumbria	\
Qualities		
Scenic views	\ * with Manchester	\
Traditional Buildings	\	X with Remote
Peace & tranquility	\	\
Wildlife	X with Manchester	X with Urban X with Remote
Family Farms	\ * with Cumbria	\ X with the Rural groups
Farming skills	\	\
Local culture	\	\ * with Urban and Rural
Social networks	\	\ * with Urban and Rural

\ no statistical difference
X statistical difference at 95% confidence
* detected preference

	Much more important				Equally important				Much more important	
Peace and tranquility		X								Wild plants, birds & animals
Upland landscapes										Community culture

Thus if a respondent thought peace and tranquillity are moderately more important than 'Wild plants, birds & animals' then they would place an 'X' as shown.

Figure 9.6   An example of grading qualities and attributes of upland landscapes (adapted from McVittie et al., 2005: 17)

Results indicate geographical qualities of the uplands (scenic views and wildlife) are considered to be more important to people living outwith upland areas. This suggests a disassociation between upland areas and the role of farming in providing public goods.

This conclusion becomes more evident when the individual attributes were statistically compared when the 'wildlife' attribute was identified as the most important of all the eight qualities, notably by the Manchester sample (Table 9.3). A range of other attributes was preferred by different sample groups. For example, the Cumbrian group had a preference for family farms over that of the Manchester sample.

In relation to the CVM exercise, respondents were initially asked their willingness to pay (WTP) for upland public goods in the UK, as a first given amount. This was then followed up with a higher or lower number depending on their response. Results were calculated using regression analysis which showed that, on average, a household was willing to pay £47 per year for upland public goods. The range was £37 to £64 at 95 per cent confidence. When the £47 is multiplied by the 24.5 million households in the UK (Office of National Statistics, 2004) a figure of £1151.5 million is reached. Within 95 per cent confidence, using the more conservative estimate of £37 the WTP drops to £906.5million. The team did, however, note that the public have a limited understanding of what constitutes an upland system as shown from the AHP results, which could have influenced the result.

Investigating the Supply Side

The supply investigation had three broad aims: to investigate the extent to which the landscape (and thereby public goods supply) is dependant on the traditional farming practices that have shaped it in the past; to investigate the importance of maintaining social capital to maintaining traditional farming practice and landscapes, and to develop an holistic conceptual model of the relationship between social capital and the environmental management of the upland regions. The link between public goods provision and social capital in the uplands is best summarised by Burton *et al.* (2005) as a combination of human capital (knowledge, skills, traditional and motivation) and true social capital (social structures and relations that enhance economic performance), which physically manifest themselves through cooperative behaviours, an example of which is the operation of the heft system as described above.

The supply side investigation used data gathered from four focus groups, eight in-depth farm family interviews and 36 semi-structured questionnaire surveys. Whilst much of the data are still under analysis, a few of more important issues will be discussed briefly here in terms of cooperative working, due to its pivotal role in social capital as a concept and, second, the provision of public goods.

Evidence from the investigations suggests that cooperative behaviour is mainly declining, particularly in terms of involvement in local communities, harvesting and shearing. Feedback from the farmers and their families suggested that this was due to the mechanisation of the manual tasks, the increased use of contractors and lower labour availability from other farms. This has led to a decline in community involvement:

When I was (young), there was the hay team and it was great fun, wasn't it? I mean, I haven't made big meals to take out into the fields for years now whereas that used to be a big part of my farm work. Cooking for everyone and taking it out into the field. (Farmer no. 1's wife)

Cooperative behaviour is however being maintained in some instances, particularly with respect to gathering on the fells and 'neighbouring' (people receiving help from neighbours). As noted above, gathering requires high numbers of people, which cannot readily be replaced by mechanisation; this is particularly the case in the more inaccessible precipitous terrain of the Lake District. In other areas all-terrain vehicles have allowed some drop in labour, but at the end of the day, the virtual boundaries of the hefts require cooperation to stop sheep scattering to other parts of the common during gathering.

With respect to public goods provision very little is known about what the public wants and how the farmers can provide it (Burton *et al.*, 2005). To help rectify this situation, the results from the AHP from McVittie *et al.* (2005) were compared to the farmers' perceptions through a simple ranking of the eight qualities (Table 9.4). This showed that wildlife was ranked more highly by the public than the farmers. Second, that the farming community (made up of four qualities) was the most important for the farmers, but least important for the public. Third, that peace and tranquillity is a perception related to the environment from which the sample is drawn and finally, traditional buildings and walls where more important to the farmers than the public.

Table 9.4   A comparison of farmers' perceptions of the benefits of upland farming with that of the public in samples from Cumbria and Manchester

	Sample Farmers	Cumbria Public	Manchester Public
Traditional farming skills	1	5	6
Small family farms	2	4	8
Strong local culture	3	2	5
Traditional buildings and stone walls	4	6	7
Wildlife	5	1	1
Community culture	6	3	3
Scenic views	7	8	4
Peace and tranquillity	8	7	2

The overall views of the farmers were that wildlife was an externality of the farming process apart from those impacting directly on farm management. Landscapes were functional to farmers for farm management reasons and were not seen as scenic views. They know the public come to seen the landscape, but they gave many examples where the public do not see the connection between landscape and farming. Furthermore, there arose a mismatch between what farmers thought the public wanted to see (tidy farms and good walls) and what the public actually wanted to see (wildlife).

In summary, the social capital project demonstrated that there was a lack of understanding on the part of both the public and the farmers as to the public goods value of the other. Whilst the public wanted landscape and they were prepared to pay for it, they failed to see it was a product of farming. For the farmers, wildlife was simply an adjunct of their operation. This still represents a productivist mentality amongst the hill farming community despite the continuing modulation of funding to Pillar II schemes within the Common Agricultural Policy.

Hill Lamb/Mutton Marketing Schemes

The main aim of the breed projects is to support upland farming families in seeking out diversification opportunities for their wool and meat products by using the unique selling properties of hill breeds.

The Swaledale Breeders Association is developing three main projects under the CHSI. The first two are designed to raise the profile of the breed amongst the public, through the creation of a Swaledale visitors centre based at Kirkby Stephen auction mart in south-east Cumbria and by demonstrating the meat can be identified as having 'specialist added-value'. The latter promotion was launched at the Ritz in London by using celebrity chefs to cook what is known as 'light lamb' (this is lamb from an animal that had not been weaned). The third project is more ambitious, the construction of a cutting and packing plant at Junction 38 of the M6 (personal communication, V. Waller, 2005). Such a scheme will allow the initial producer of the meat to obtain higher sales prices and reduce food miles. This enterprise has 40 farmers and is one of the few true cooperative farmer ventures in the county. It is funded by LEADER+, Distinctly Cumbrian[7] and Rural Regeneration Cumbria.[8]

The Rough Fell Sheep Breeders Association (RFSBA) is aiming to promote the breed producers, their products and the link to the landscape and heritage of the breed's indigenous area. In order to achieve this aim the Association has already produced a promotional video entitled '*Rough Fell Heritage – a celebration of the life, work and landscape of the Rough Fell sheep Farming Community*'. They are also developing a collaborative project which enables cooperative working to supply meat direct to commodity buyers in the mainstream food sector. As a breed with a limited flock size, this will enable them to have a greater selling power, not unlike the past Marketing Boards in the 20th century (Ilbery, 1984).

A particularly important project the RFSBA are working towards is the labelling of their meat as a *Product of Designated Origin* (PDO) or *Product of Geographical Indication* (PGI). These are EU initiatives which allow for the creation of a protected food name, such as that of Parma Ham. The difference between the two labels is that a PDO requires the product to be produced, slaughtered, cut and processed completely within the region of origin. For the PGI, the area must have some recognised historical character, and the product must be grown and processed in the region. Thus the regulations for obtaining a PDO are stricter than a PGI (personal communication, A. Banford, 2005). For the Rough Fell Breed it is hoped a PDO can be achieved due to the location of a slaughter house at the southern end of the breed's indigenous area.

The development of the Herdwick brand is probably the most well known of upland breed diversification schemes in the country. Championed by the Prince of Wales, the Herdwick Sheep Breeders Association (HSBA) boasts many diverse product lines including: Lakeland Herdwick Direct (meat); a partnership with Goodacre's Carpets of Kendal Ltd producing Herdwick carpet; and the production of premium fleeces for craft spinning, knitting and felt making by a number of cottage industries (www.herdwick-sheep.com 2005). The HSBA are seeking a PDO as well; however, they have a problem due to the lack of slaughter facilities within their delineated region. The only slaughter house is a low throughput facility on the Furness Peninsula which specialises in rare breeds and Herdwicks. If the facility was expanded, the financial costs would be difficult to sustain, as this would push the premises into the size bracket where a vet and a Meat and Livestock Commission officer would have to be permanently employed on site (personal communication, A. Banford, 2005).

Discussion

The complexity of the sector's economic and social decline and its impact on the environment in the 20th century followed by the shift to post-productivist aims has created many issues for the hill farming industry. Two common themes that have emerged are: the continuation of the industry *per se* (including: economic viability, succession and non-productivism) and the maintenance of the upland landscape in a form desired by general public. The three-pronged approach developed by the CHSI through research, vocational and economic projects has gone some way to address the complexity of these two challenges.

The lack of succession and labour shortages facing the hill farming sector was demonstrated by the Social Capital project through the deterioration of cooperation between farms over time. Gathering is probably the last real area of cooperation, due to management necessity. However, even this is suffering as the collapse in paid farm labour had reduced the number of people available. The CHSI has responded to this challenge via the FFTS. Farmers have realised through this project that alternative employment mechanisms, such as fractional shared labour, can assist the situation, whilst at the same time it demonstrated that there were young people interested in starting in the industry.

As with all pilot initiatives, the FFTS suffered from teething problems particularly in relation to the underpinning knowledge of the trainees, their awareness of alternative funding (especially agri-environmental initiatives) and transportation logistics for younger trainees. The poor responses by the trainees for the environmental goods exercise here was interesting, as it corroborated the lower importance placed on wildlife by farmers involved in the supply side research of the Social Capital project. Despite this, all the farmers involved in both the FFTS and the SCHF projects had entered into either ESA or CSS agreements (Mansfield and Martin, 2003; Burton *et al.*, 2005). This divergence between the cognitive and the conative (behaviour) suggests that the farmers have diversified into these environmental grants for financial reasons and not really for the benefit of the landscape, wildlife and public. To the Cumbrian farmers in the projects, landscape and wildlife enhancement schemes are just *one possible* solution to overcoming their decreasing gross margins of the farm business. It would seem, therefore, that the farmers have adopted these agri-environment initiatives within their own terms of reference, rather than those espoused by the Government and the public.

Given the close relationship between the trainees of the FFTS and the farmers they worked with, it would seem that the view that the industry is productivist rather than post-productivist has also been perpetuated to the next generation. This attitude is further demonstrated through the evidence of farm diversification in the SCHF project, whereby the non-farming operation was often managed by the partner (Burton *et al.*, 2005). In this way the farmer has been able to focus on their stock management role, re-enforcing the 'productivist' culture. Unfortunately, for the farmer, the shift to Pillar II articulated by the Rural Development Regulation has meant that a wedge is being driven between what they traditionally know and practice (that of livestock production) and that which is supported (ie. non-farm on-farm diversification including alternative enterprises, tourism and environmental initiatives).

In a way, the Hill Lamb and Mutton Initiative can be viewed as a mechanism to bridge the void between the productivist traditional hill farmer and the post-productivist vision of EU and UK agricultural policy, by maintaining the need for stock production, and at the same time diversifying the farm business away from purely meat and wool. Whilst the breeder associations have made substantial developments with respect to this need, new issues have emerged. The largest challenge, that of availability of local slaughter facilities, is rapidly becoming the biggest barrier to success by some of the projects within each breeder group. The other main issue facing these native breed schemes is the problem of *single dale dominance* by first innovators. The character of the upland farming landscape means that farmsteads are usually located close to each other in the valley bottom. The sparse distribution and clustered nature of the farmsteads brings farm businesses into direct competition with each other for diversification opportunities. Under traditional adoption theory, a new idea is taken up by an individual and then the idea spreads to the neighbours to form the early and late majority, and finally to the laggards, who may or may not adopt (Rogers, 1962). Unfortunately the geography of upland farms means that once the innovator has begun to operate, the surrounding farms have no chance to follow suit as the local market is rapidly saturated. This has happened in Borrowdale in the Lake District and around Kirkby Lonsdale in the extreme south of the county. The only time a neighbour can adopt the idea is if the initial innovator's business gets big enough to expand out of the local area into the wider national market place, leaving a void in the original local market.

Compounding the economic situation is the debate related to public goods. Through the SCHF results it is apparent that there is a difference of opinion as to the value of the uplands between farmers and the public. Whilst wildlife is seen as the most important element of the upland landscape and the public were willing to pay for it, there was no real recognition that it was agricultural activity that had produced it. This was particularly true for the urban sample, who are no longer intimately connected with the act of food production. The rural sample seemed to have more of a connection with the farmed landscape; this is probably a function of village life in Cumbria, where farming families are still (just) an integral part of the rural community. This lack of recognition, in a perverse way, underpins the aims of the post-productivist vision that the effect and not the cause of the upland landscape in Cumbria is desired.

Conclusion

Whilst the CHSI has gone someway to exploring and addressing the twin issues of continuation of the hill farming industry and the maintenance of the upland landscape desired by the public, it has uncovered a more fundamental challenge for rural development policies and initiatives in Cumbria and beyond. The issue is that the farmers' traditional products are no longer wanted but instead their by-products are. For the hill farmers, the task is to adapt and accept that currently a non-productivist countryside is desired instead of a productivist one. This is the challenge.

At the same time, there needs to be formal recognition, through policy instruments, that the intangible public goods of the upland landscape are only there because of the hill farming, and that the tradition and skills of the industry are required to achieve this. To address this conundrum one such solution for rural development policy makers is to use hill farming subsidies to pay for public goods rather than meat and wool, which is really what the Hill Farm Allowance is now doing (DEFRA, 2005). In this way the underpinning philosophy should therefore be that payments are for farm management and related livestock production, but the product is landscape and wildlife. If this idea is taken to its natural conclusion, then the physical manifestation of policy should be developed that pays people to *be* hill farmers, much like we pay for people to *be* countryside managers. In the countryside management industry the landscape and wildlife are the products and any by-products simply generate extra income, for example, charcoal from woodland management. Therefore we have moved from a situation where environmental goods are the by-product of farming to a situation where farming is by-product of environmental goods. The hill farming industry thus moves from being a primary producer to being a tertiary provider. The CHSI has, therefore, provided a step towards the evolution of a new hill farming industry for the 21st century.

Acknowledgements

The author would like to thank the contributions of Adrian Banford, Veronica Waller and Hilary Wilson in the clarification of some of the more complex issues covered in this chapter. The author would also like to acknowledge the funding for the FFTS evaluation and the SCHF through LEADER+ and the International Centre for the Uplands respectively, and the collaboration with her colleagues on both projects.

Notes

1 The walls are not mortared together but are constructed as two separate uncemented walls, tied together with stones crossing from side to side and the gap in-filled in between with smaller pieces. Walls last for about 120 years, but the actual lines of many walls have existed for hundreds if not thousands of years and are simply re-built when they collapse.

2 The Farmers sample was taken from 3 upland areas of Cumbria surveyed as part of the 'supply' side of the Social Capital in Hill Farming project. The supply side investigation used data gathered from four focus groups, eight in-depth farm family interviews and 36 semi-structured questionnaire surveys.

3 Voluntary Action Cumbria – this organisation is the Rural Community Council for the county of Cumbria which is funded through a variety of mechanisms to provide innovative methods of rural development and supply financial aid. Much of the financial aid is channelled through recognisable funds such as LEADER+, the Countryside Agency and the County Council.

4 The 2 objectives of UK LEADER+ are: first, to build capacity in local rural communities to encourage development of potential, and second, to support local communities to develop and implement high quality innovative strategies which identify new ways of:

   a) Creating new jobs and increasing economic opportunity
   b) Improve the quality of life
   c) Protect and enhance natural and cultural heritage
   d) Improve the organisational skills of rural communities.

5 The organisations involved include: LEADER+ under the auspices of Voluntary Action Cumbria, the Lake District National Park, The National Trust, The Yorkshire Dales National Park, Friends of the Lake District; English Nature; National Farmers Union.

6 The National Trust is a charity set up in 1895 to act as guardian for the nation in the acquisition and protection of threatened coastline, countryside and buildings. Currently it owns 45,000 ha and 91 farms (National Trust, 2005) in the Lake District National Park, which makes up the central massif of Cumbria (Figure 10.3).

7 Distinctly Cumbrian is an organisation that promotes Cumbrian-made Products.

8 Rural Regeneration Cumbria – the Government project set up in the aftermath of Foot and Mouth to help the Cumbrian rural economy reinvigorate.

References

Aitchison, J., Crowther, K., Ashby, M. and Redgrave, L. (2000) *The Common Lands of England: a biological survey. County Report for Cumbria.* Aberystwyth: Rural Surveys Research Unit. Contract for DEFRA.

Backshall, J. (1999) Managing bracken in the English Uplands. *Enact*, 7(2): 7–9.

Backshall, J., Manley, J. and Rebane, M. (2001) *The Upland Management Handbook.* Peterborough: English Nature.

Burton R., Mansfield, L., Schwarz, G., Brown, K. and Convery, I. (2005) *Social Capital in Hill Farming.* Hackthorpe: Report for the International Centre for the Uplands.

Chadwick, L. (2003) *The Farm Management Handbook 2002/3.* Ayr: Scottish Agricultural College.

Chitty, G. (2002) *Study of Cultural Landscape Significance: proposed Lake District World Heritage Site.* Carnforth: Hawkshead Archaeology & Conservation.

Cumbria County Council (1997) *State of the Environment Audit.* Carlisle: Cumbria County Council.

Department of Environment, Food and Rural Affairs (DEFRA) (2004) *Agriculture in the UK 2003.* London: The Stationery Office.

DEFRA (2005) *Environmental Stewardship.* London: Rural Development Service.

English Nature (2001) *State of nature: the upland challenge.* Peterborough: English Nature.

Franks, J., Lowe P., Phillips, J. and Scott, C. (2003) The impact of foot and mouth disease on farm businesses in Cumbria. *Land Use Policy*, 20: 159–168.

Gray, J. (1998) Family farms in the Scottish Borders: a practical definition by hill sheep farmers. *Journal of Rural Studies*, 14(3): 341–356.

Grigg, D. (1995) *An Introduction to Agricultural Geography.* London: Routledge, second edition.

Hart, E. (2004) *The Practice of Hefting.* Ludlow: E. Hart.

Haskins, C. (2001) *Taskforce for the Hills – Report to the Minister for Agriculture.* London: DEFRA.

Ilbery, B.W. (1984) *The Geography of Agriculture: a social and economic analysis.* Oxford: Longman.

Ilbery, B.W. and Bowler, I.R. (1998) From productivism to post-productivism. In Ilbery, B.W. (ed.), *The Geography of Rural Change.* London: Longman, pp. 57–84.

LEADER+ (2001) *Cumbria Fells & Dales Leader+ Programme – Development Plan.* LEADER+, Cumbria, unpublished.

Lowe, P., Buller, H. and Ward, N. (2002) Setting the next agenda? British and French approaches to the second pillar of the Common Agricultural Policy. *Journal of Rural Studies*, 18: 1–17.

McVittie, A., Moran, D., Smyth, K. and Hall, C. (2005) *Measuring public preferences for the uplands.* Hackthorpe: Final Report to the Centre for the Uplands.

Mansfield, L. and Martin, H. (2004) *The Fell Farming Traineeship Scheme: Final Evaluation.* Cumbria: On behalf of Cumbria Fells & Dales LEADER+.

Ministry of Agriculture, Fisheries and Food (MAFF) (2000) *Agenda 2000: Annex 3.* London: MAFF.

Office of National Statistics (2004) *Social Trends 34.* London: Stationery Office Books.

Ratcliffe, D. (2002) *Lakeland: the Wildlife of Cumbria.* London: HarperCollins.

Rogers, E.M. (1962) *The Diffusion of Innovations.* New York: Macmillan.

Thompson, D.B.A., MacDonald, A.J., Marsden, J.H. and Galbraith, A. (1995) Upland heather moorland in Great Britain: a review of the international importance, vegetation change and some objectives for nature conservation. *Biological Conservation*, 71(2): 163–178.

www.herdwick-sheep.com (2005) Herdwick Sheep Breeders Association website, accessed 18 July 2005.

Chapter 10

Participation and Stewardship: Sustainability in Two Canadian Environmental Programmes

Guy M. Robinson

One of the central characteristics of attempts to promote sustainability as part of programmes aimed at delivering environmental benefits has been the incorporation of community participation. This can take various forms, but has often been closely linked to schemes with a strong 'bottom up' dimension as opposed to government-directed 'top down' control. This chapter examines two particular environmental programmes in Canada, one in the Atlantic provinces and one in Ontario, with a view to assessing how the process of 'bottom up' planning has worked in these two instances and what lessons may be learned about the long-term sustainability of the actions incorporated within the two case study schemes.

Canada's Atlantic Coastal Action Program (ACAP)

The Atlantic Coastal Action Program (ACAP) is one of five regional ecosystem-based initiatives developed by Environment Canada in the 1990s as best examples of how sustainable development might be achieved through a combination of economic, environmental and community-based components (Environment Canada, 1994; 1995). The aim was to use natural geographic units such as watersheds, within which Environment Canada could work with various partners to produce a series of tangible sustainable outputs. Community involvement has been a central element within these initiatives, being especially strongly emphasised within ACAP, which was launched in 1991 with Cn$6 million from the federal government's Green Plan. The basic aim was to restore and maintain environmentally-degraded harbours and estuaries in areas where significant pollution problems are being experienced. Thirteen areas were selected to receive funding, a strong factor in selection being the willingness of key local stakeholders to participate in round table discussions that would determine the nature of ensuing action plans. The areal extent of the 13 areas was variable so that drainage basins as well as estuaries and coastlines were included and so that involvement from a range of economic interests could occur, notably agriculture, pulp-milling, extractive industry and fishing. Settings varied from urban areas with heavy pollution of harbours (e.g. St. John's and Saint John) to traditional industrial centres with long-term environmental degradation (e.g. Pictou, Sydney, Corner Brook) and agricultural areas with run-off from heavily fertilised and chemically-treated farmland (e.g. Bedeque Bay, Cardigan Bay and the Annapolis River). A fourteenth area, the Sable Island Protection Trust, was added in 1999, and as part of a third phase of the Program, two further areas in Labrador were added in 2005 (Southcoast Labrador/Gilbert Bay and Happy Valley-Goose Bay) (Figure 10.1).

The diversity of the 16 areas has produced variety in the foci and approaches of the individual management committees. Each of the 13 committees was charged with the production of a comprehensive environmental management plan (CEMP) at the end of an initial six-year funding period. Each committee received an initial allocation of Cn$50,000 per annum, partly to support the hiring of a coordinator. The committees have acted as management boards, each receiving input from Environment Canada but functioning as partnerships between stakeholders representing local municipalities, industries, local businesses, community groups, academia, environmental groups, and provincial and federal government. The breadth of membership of the committees has meant that the degree of control exerted by

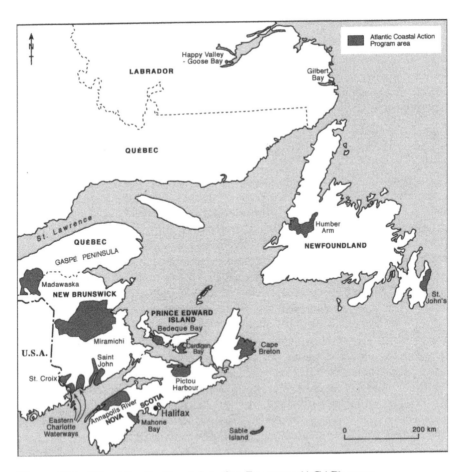

Figure 10.1   The Atlantic Coastal Action Program (ACAP) areas

Environment Canada has often been relatively limited, evolving into an advisory capacity through its representative on each committee. Indeed the agency does not formally approve work-plans developed by the committees, although it allocates financial support and advice on the basis of its assessment of the plan.

The initial agenda for each individual ACAP committee was a seven-point programme devised by Environment Canada (Table 10.1) and five principal areas of concern to be addressed: sustainable livelihoods, natural heritage, water quality, responsible stewardship and ecosystem planning. However, the diversity of the ACAP groups has been shown in their individual responses to these five broad concerns. The groups have made their own determinations of the priority areas in their own areas, as expressed in the various CEMPs.

Essentially the CEMPs were intended to lead "to a thorough investigation of the critical issues affecting local resources, an assessment of the remedial options available to the community, and a choice of options which best served the primary environmental, and in some cases, socio-economic objectives of the community" (McNeil et al., 2006: 370). The CEMPs have acted as 'roadmaps' for the work of the ACAP groups, and they have been revisited and updated regularly. Whilst the CEMPs were being formulated certain preliminary projects were implemented, including monitoring activities and a variety of demonstration projects to raise visibility within the community. Hence there has been nearly 15 years in which various environmental actions have been taken as part of ACAP. By the end of 2005, over 8000 individual projects had been implemented. Table 10.2 provides a summary of the first decade of ACAP projects, in which 44 per cent of projects were classified by Environment Canada as comprising part of their 'Nature' priority, focusing on water quality and protection, enhancement of wildlife and riparian habitats and protecting sensitive environments. One-third of projects concentrated on 'Clean Environments' through monitoring activities, restoration of contaminated lands and

Table 10.1   Environment Canada's Seven-Point Programme for each ACAP Group

- Appointment of a full-time community co-ordinator and office for each of the project sites,
- Assessments of environmental quality, including identification of all sources of environmental problems,
- Development of a long-term vision, supported by clear objectives necessary to obtain the long-term goals,
- Identification and assessment of necessary remedial actions and conservation measures,
- Development of comprehensive environmental management plans,
- Promotion of environmental stewardship through education and awareness activities,
- Implementation of pilot projects that will demonstrate the importance and effectiveness of low-cost, innovative solutions to environmental issues in watersheds.

water conservation programs. Around 11 per cent of projects were aimed at 'Climate Change and Air Quality', focusing on reducing greenhouse gas emissions and developing energy conservation measures. The remaining 12 per cent contributed to 'Management and Business' by working with firms and government organisations to raise environmental awareness.

The notion of the community as the basic building block for socio-economic development has gained greater currency over the last four decades, following negative experiences with 'top down' development projects in developing countries in the 1950s and 1960s. This top-down approach, determined by national governments and large non-governmental organisations such as the World Bank, tended to be characterised by narrow economic principles that neglected local, self-determined objectives (Friedmann and Weaver, 1979). An alternative, as partially exemplified in the ACAP scheme, is to maximise the mobilisation of an area's natural, human and institutional resources with policies that are motivated and controlled initially 'from the bottom', that is with a community base (and sometimes termed 'integrated rural development'). At the heart of bottom-up initiatives is a reliance on local initiatives, emphasised within a range of associated terms: basic-needs oriented, labour-intensive, small-scale, regional-resource based, appropriate technology (Stohr and Taylor, 1981). From the late 1980s an additional element within these bottom-up approaches has been 'sustainable development', emphasising the twin goals of economic self-sufficiency and production for limited use (Redclift, 1992). The surge in popularity of 'New Right' governments in the Developed World in the late 1980s helped promote the 'self-reliance' component of local initiatives (Green, 1987) and stimulated an outpouring of research analysing the background and effectiveness of the various approaches to bottom-up, community-led planning and development (Wilkinson, 1991).

A recurrent theme in this research was the variety within so-called community-based schemes: from complete control by self-regulating citizens' groups to a large element of control still within the hands of local or even state government. Hence, 'bottom-up' may seem to be an inappropriate term in those cases where the initiative for a particular programme resides within an arm of government despite strong community-level involvement and even direction. In some respects ACAP is a fairly typical example of the type of collaboration that can occur between communities and a government agency, in which cooperation between community and government is viewed (ideally) by both sides as a satisfactory means of tackling a particular set of problems.

One of the key elements within ACAP is the multi-partner round table or board. These boards, comprising stakeholders or 'interested parties' have tended to be interpreted in terms of representatives of sectoral interests such as farming, fishing, pulp-milling and government. Representation on individual ACAP boards by members of the public, as opposed to those from specific sectors, has been more limited. This prompts questions about the extent of community involvement in the scheme and of whether the agenda pursued by the boards reflects community views or a more narrow set of ideas emanating from particular sectoral interests. This problem has been raised in the context of other 'community-based' schemes, highlighting "the danger ... of inferring community involvement on the basis of the

Table 10.2   Key outcomes from the first decade of ACAP schemes

Environment Canada Priority	Results
Nature – includes installing in-stream structures to improve habitat, riparian zone re-vegetation, creating artificial wetlands, and protecting environmentally sensitive areas	• enhanced 5 sq km of wildlife habitat • installed 261 in-stream habitat structures • planted 60,547 native plants, trees and shrubs • protected 182 sq km of riparian habitat
Clean Environments – monitoring water quality in watersheds, cleaning up contaminated areas, addressing sewage treatment issues, carrying out water conservation programs in homes and businesses	• cleaned up 65 illegal dumpsites • installed 3 pump-out stations for boat sewage, eliminating 82,740 litres of raw sewage from the coastal environment • securing funding to develop 4 municipal sewage treatment plants (St John's, Lunenburg, Miramichi [2]) • diverted 538 tonnes of waste from landfill • installed 677 water-saving devices • conserved 33,404,810 litres of water • eliminated 121,818 litres of paint from the waste stream • eliminated 33,000 mg of mercury from the waste stream • 22 sq km of shellfish harvesting opened up (which had previously been closed by pollution)
Climate Change and Air Quality – includes projects to reduce Greenhouse gas emissions (e.g. household energy audits, vehicle emissions clinics) and working with business and industry sectors on energy conservation programs	• performed 461 energy audits • reduced greenhouse gas emissions by 30,000 tonnes • tested 1,232 cars at vehicle emissions clinics • conserved 42,466 kwh of energy

Source: McNeil et al., 2006: 372.

participation of a small number of people not necessarily representative of wider local views" (Shortall, 1994).

The author's random sample survey of 100 households in three of the more rural ACAP areas (Annapolis Valley, Bedeque Bay and St Croix) revealed quite strong polarisation within the communities with respect to awareness of the scheme and active participation (Robinson, 1997a). Hence 56 per cent of respondents to a questionnaire survey had no or very little knowledge of the scheme and 82 per cent had no involvement with it despite the fact that, amongst the respondents, 82 per cent had been resident in the local community for over five years, and 56 per cent for over 20 years. In contrast, 44 per cent of respondents displayed significant levels of awareness, which may be seen perhaps as a first step towards more broadly-

**Table 10.3   Key outcomes of the activities of the Clean Annapolis
                River Project**

---

1. Trained over 300 volunteers in water quality monitoring.
2. Restored 10 000 metres of fish habitat.
3. Over 500 presentations to various schools and community groups.
4. Created approximately 60 person years of environmental industry employment in the local area.
5. Volunteer River Guardians have collected over 1600 water samples.
6. Planted over 5000 trees in riparian zones.
7. Carried out water conservation throughout the region.
8. Distributed 10 000 booklets on onsite septic system care and management.
9. Signed stewardship agreements on approximately 20 hectares of salt marsh.
10. Involved in the construction of approximately 100 hectares of wetland.
11. Erected over 100 nest boxes in wetland areas.
12. Identified ground level ozone as a problem in the Annapolis Valley.
13. Completed mapping showing areas at risk for coastal flooding.
14. Researched the impacts of climate change in the watershed.
15. $3,000,000 direct spending in the local economy, and economic spin-off.

Source: CARP, 2006, *Clean Annapolis River Project*, http://www.annapolisriver.ca/mission.htm.

based community involvement. Direct participation of individuals in the scheme varied quite considerably between groups, reflecting the degree to which the ACAP groups had developed participatory initiatives. For example, both the St. Croix and Annapolis Valley groups had developed water quality monitoring activities utilising volunteers. The latter group had instigated over 50 Annapolis River Guardians operating monitoring at 40 sites throughout the drainage basin and providing data for local scientific organisations. The targeting of these participatory activities at school children and students aims to raise environmental awareness amongst the younger age groups in the community as part of the establishment of a stronger aspect of environmental citizenship. This is a tangible positive outcome of the scheme.

There are other notable gains associated with ACAP, principally those related to the income generated for environmental projects and measurable environmental benefits produced from ACAP-related activities. In the first five years of its operation, in return for government input of Cn$6 million, in excess of Cn$30 million was generated from other sources. In ACAP areas, such as the Annapolis Valley, where the board has been especially entrepreneurial, the additional income generated has enabled environmental improvements that have been far more wide ranging than might at first be expected from the Program outline which emphasised the 'coastal environment' (see Table 10.3). Over 90 volunteers have been involved in the collection and analysis of over 3500 water samples in its fourteen-year history. Quality assurance issues have been addressed utilising procedures developed by community groups in the United States and incorporating input from Environment

Canada and Acadia University. This community-based monitoring has direct benefits to government "through the extension of their monitoring networks, cost savings, promotion of public participation to achieve government goals, and providing an early warning system of ecological changes" (Sharpe and Conrad, 2006: 396). For local communities there is a development of social capital and opportunities for direct input into the management of natural resources. Nevertheless, a review by the Clean Annapolis River Program (CARP) of its River Guardians program concluded that it had produced little direct impact on local policy and decision-making and that the public profile of the program had remained at a low level despite its longevity (Sharpe and Sullivan, 2004).

The combination of 'top-down' initiation from Environment Canada and 'bottom up' leadership from the ACAP co-ordinators and management boards has forged a new approach to tackling complex problems in coastal, estuarine and riparian environments of Atlantic Canada. The collaboration between key stakeholders represents a considerable advance on the previous degree of involvement in environmental protection and management by individuals from the main sectors of community life. In particular, it has helped industrial leaders to become more aware of outside views of their activities whilst also enabling industrial participation in schemes to ameliorate pollution.

From 1997 an ACAP Science Linkages initiative has endeavoured to extend partnerships between ACAP groups and scientists from Environment Canada so that there can be more scientific input to ACAP activities. Between 1997 and 2003 this led to Environment Canada investing over Cn$1 million to fund 95 projects, which raised an additional Cn$4.5 million (Dech, 2002). This Initiative ensures that ACAP contributes directly or indirectly to Environment Canada's priorities. Overall, a further Cn$6 million from Environment Canada was invested in ACAP between 1997 and 2001. The agency calculated that between 1997 and 2001 ACAP-related activities produced Cn$22 million in direct and spin-off economic activity. Expenditures on administration of ACAP projects generated Cn$4.4 million in federal and Cn$3.6 million in provincial tax revenue (Environment Canada, 2002). As part of this funding Environment Canada has encouraged ACAP groups to work as part of larger ecosystem-based coalitions such as those representing the Gulf of Maine, the Bay of Fundy and the Southern Gulf of St. Lawrence. Additional provincial programmes include the Nova Scotia Sustainable Communities Initiative, which focuses on communities along the Fundy coast and around the Bras d'Or lakes in Cape Breton.

In 2005 ACAP was extended for a further five years, with funding tied to effective delivery and maintenance of the community-oriented approach, was agreed jointly by Environment Canada and the individual ACAP boards. The five theme areas depicted in Table 10.4 have been selected as the key foci for this third stage of the scheme. More accountability has been introduced through the initiation of a results-based management and accountability framework (RMAF), which will evaluate outcomes from the various ACAP projects against a set of pre-determined benchmarks including performance measurement, evaluation and reporting strategies. Created by Gardner Pinfold Consulting Economists (GPCE), who evaluated the economic impacts of ACAP, the use of a RMAF ensures that ACAP complies with the standards set by

Table 10.4   Environment Canada's Plans for ACAP (2004–2009)

1. Continue our highly effective and successful working relationship with the 14 ACAP organizations[1] through knowledge generation, capacity building, collaborative science and action, with an emphasis on environmental results;
2. Expand, as capacity allows, the successful approach into new sites (e.g., Labrador), through the adoption or mentoring of adjacent watersheds, and into coastal areas where appropriate (e.g., integrating with the Department of Fisheries and Oceans' coastal and ocean agenda);
3. Continue to work with and build the capacity of multi-stakeholder coalitions organized around larger regional ecosystems (e.g., Bay of Fundy Ecosystem Partnership, Southern Gulf of St. Lawrence Coalition on Sustainability, Gulf of Maine Council, and others to be developed) based on 'ACAP principles';
4. Build and maintain our 'Community of Practice' through the continuation and broader application of initiatives such as the Annual ACAP workshop/ gathering, meetings of the Coordinators, the ACAP Advisory Council, site-to-site and government-community exchanges, theme-based collaborative projects, Community Bulletin Board, etc.; and
5. Strengthening inter-departmental and inter-governmental collaboration in support of communities, through the ongoing development of Sustainable Communities Initiatives (as currently being piloted as the Nova Scotia Sustainable Community Initiative).

[1]   A further two ACAP areas have since been added in Labrador.

Source: Environment Canada, 2004.

the Treasury Board of Canada for projects receiving public funding. GPCE's report found that it would have cost Environment Canada twelve times the current ACAP budget to deliver directly the same outputs as the ACAP groups over the period 1997 to 2002. During this time ACAP created 482 person-years of employment and generated taxable revenues of CN$8 million to provincial and federal governments (McNeil *et al.*, 2006).

In addition to the measurable economic, social and environmental impacts of ACAP there are more intangible benefits related to community development and to changes in the nature of the relationship between communities and the government agency, Environment Canada. Community-based action has been shown to produce positive environmental benefits that have also possessed economic and social multipliers. Moreover, by transferring certain responsibilities and actions to the community not only has Environment Canada been able to successfully implement certain desirable actions cost effectively but it has also produced outcomes that would have been impossible under a simple 'top down' delivery system. One of the reasons for the latter is the diminution of the 'us and them' adversarial role often associated with the 'top down' model in which government is frequently viewed as primarily an enforcer of environmental regulations rather than as a partner in a collaborative network that is striving for environmental improvements. This removal of the adversarial approach has led to widespread involvement by the commercial sector in the individual ACAP boards. Particularly noteworthy has been the role of representatives of the local pulp and paper mill, Neenah Paper, in the Pictou Harbour Environmental Protection Project, where the mill now operates a state-of-the-art wastewater treatment system that more than matches federal environmental effluent regulations (McNeil et al., 2006: 377).

McNeil et al (2006) argue that ACAP has had significant impacts at a variety of spatial scales, from the local to the international. They claim that environmental awareness has been greatly raised at local level, for example through farmers signing up to agri-environmental projects within ACAP and then passing on their environmental knowledge to the non-farming community via farm-based educational visits. At regional and national levels it has been some of the urban-based ACAP groups that have had a major impact through the development of improved sewage management to reduce pollution of harbours and revealing the extent of riparian and estuarine pollution through monitoring of water quality. At international level there has been widespread transfer of knowledge to other countries interested in promoting community-based environmental management and also the development of effective partnerships on both sides of the Canada-US border to tackle environmental issues affecting the whole of the Bay of Fundy region.

The Ontario Environmental Farm Plan (EFP)

The second Canadian example of a bottom-up environmental scheme to be discussed here is an agri-environmental scheme formally launched in 1993, the Ontario Environmental Farm Plan (EFP). Formulated by a grouping of Ontario farm organisations, the Ontario Farm Environmental Coalition (OFEC), this set an agenda for the province's farming community to adopt with respect to environmental concerns associated with agricultural production practices. Based on an earlier project in Kansas (Castelnuovo, 1999), the scheme identified key environmental issues relating to water quality (with special reference to agricultural nutrients), soil quality, air quality, agricultural inputs and natural areas such as wetlands and woodlots. It recommended that every Ontario farmer should develop and implement an environmental farm plan designed to meet the needs of their own farm. This was then translated into a series of worksheets for participating farmers to complete, under the headings of farm productivity and profitability, aquifer degradation, surface water degradation, farm health and safety, and air pollution. The main issues addressed in the EFP are indicated in Table 10.5.

Funded through Agriculture and Agri-Food Canada's Green Plan Program, the EFP provides up to Cn$1500 per farm business to help farmers implement new management practices to address environmental problems. An additional incentive is small prizes for the best individual plans produced each year. Farmers entering the scheme do so voluntarily and follow a six-stage process (Table 10.6), from attendance at an introductory workshop to full implementation of a plan based on completion of approved worksheets. Each farmer prepares a plan for their own farm,

Table 10.5   Worksheets in the Ontario Environmental Farm Plan

Soil and site evaluation	Water wells
Soil management	Pesticide storage
Nutrient management in growing crops	Fertiliser storage
Manure use and management	Petroleum products storage
Field crop management	Disposal of farm wastes
Pest management	Treatment of household wastewater
Stream, ditch and floodplain management	Storage of agricultural wastes
Wetlands and wildlife ponds	Livestock yards
Woodlands and wildlife	Silage storage
Energy efficiency	Milking centre wash water
Water efficiency	Noise and odour
Horticultural production	

Source: OFEC (1999)

which is then approved by a local peer review committee consisting of members of the local farming community, farmers' organisations and possibly a representative from the provincial agricultural ministry. In completing the worksheets farmers are encouraged to pursue actions to minimise environmental risks (Bidgood, 1994), highlighting areas of environmental concern and setting goals for improvement to an agreed timetable. In evaluating any given environmental issue, a four-point scale is used (from 4 = conditions already protecting the environment/carry least potential for environmental damage; to 1 = conditions with the highest potential to affect the environment).

The worksheets were designed by over 100 individuals working cooperatively through 23 committees consisting of farmers and both governmental and non-governmental representatives of farming organisations. Hence there was a collaboration between government and the farming community in designing this agri-environment scheme, and also in its subsequent implementation. However, much of the control over the environmental actions taken rests with individual farmers and farmers' organisations.

During the first decade of its operation Cn$48 million were spent by government on on-farm improvements through the EFP. Moreover, for every dollar received for support of measures in the EFP, it is estimated that farmers spent 3 or 4 more dollars from their own pocket (Klupfel, 1998: 16). This expenditure represented around 40 per cent of the province's farmers participating in introductory workshops and just under one-quarter of farmers (and one-fifth of farmland) proceeding through all six stages. Data from the Ontario Soil and Crop Improvement Association (OSCIA)

Table 10.6   The six-stage sequence of the EFP

Stage	Actions
1. Introductory workshop	site evaluation; assess potential concerns
2. Complete farm review	review farm operations; complete relevant worksheets
3. Second workshop	consider possible actions; learn how to develop a realistic plan
4. Complete action plan	identify actions for all 'Fair' or 'Poor' rated situations; develop timetable for action
5. Peer review	add suggestions/ask for changes; return plan to farmer; send information anonymously to OFEC
6. Implementation of plan	put plan into action; re-evaluate each year

Source: OFEC (1999).

show a concentration of participants in the province's main livestock production areas (eastern and central Ontario) and much lower participation in areas of fruit and vegetable production (e.g. Niagara). The smallest and very largest farms are those least likely to participate (Robinson, 2006a; 2006b).

Nearly 10,000 worksheets were completed during this first decade of the Plan. Data on these from the OSCIA indicate the wide range of environmental actions covered in the individual EFPs. Soil management activities accounted for roughly one-quarter of all actions. This included application of conservation tillage to reduce soil erosion, greater use of crop rotations, planting of cover crops and shelter belts, introduction of tile drainage, careful use of farm machinery to avoid soil compaction and application of cropping practices that retain a high amount of organic matter. Around one-quarter of actions were related to water management and a similar proportion to farm waste. Over 1,600 upgrades were made to wells supplying drinking water whilst concern over the handling of wastes on farms reflected the attention given to this issue following an incident at Walkerton (150 km north-west of Toronto) in summer 2000 when seven deaths and illness in over 2,300 people were caused by contamination of water supplies to the town by manure-based run-off from farmland (O'Connor, 2002). The potential for nuisance suits against farmers was also raised by this incident (Caldwell, 2001). Specific issues being addressed in the EFP include the land base available for disposal of manure and nutrient components, types of manure storage, distance from manure storage to non-farm land uses, methods of manure disposal, and the size and type of livestock operations (Henderson, 1998).

Surveys of participants in the scheme revealed various positive responses to the EFP in terms of its influence on increasing awareness of farm conservation issues, its educational value and the identification of potential environmental risks. A survey by

Fitzgibbon et al (2000) reported that just over half the farmers sampled had revised or updated their action plans and nearly two-thirds had undertaken environmental actions in addition to those stated in their own EFPs. Actions directly related to the EFPs were estimated as costing Cn$8,589 per farm and involved capital investment in new 'environmentally friendly' equipment, tile drainage and installing no-till systems.

The author's own survey in 2002 showed that, for participating farmers, its attractions were its voluntary nature, the opportunity to take desirable actions without government enforcement and input from the local community in providing guidance on conservation actions. However, as over half of the province's farmers had not yet attended EFP workshops, there remained a strong level of non-involvement that suggests barriers to participation are prevalent (Smithers and Furman, 2003). The author's interviews with farmers indicated that key barriers were perceived costs of environmental actions and farmers' personal priorities, in which environmental issues were often deemed as relatively unimportant. A key aspect was the extent to which the farmer perceived participation in the scheme as being a central element in their normal farming tasks. Many of those signing up for the EFP regarded the ensuing actions as representing 'good stewardship', and hence they wished to participate so that their peers would regard them as good stewards of the land. Non-participants tended to regard the measures that are integral to the EFP as either 'extras' that did not really apply to them or as simply unnecessary or even as things that they operate already without needing to enrol in a special plan. The latter sentiment was frequently expressed by fruit and vegetable producers.

A key difference between the measures supported by the EFP and those in most European agri-environment schemes is that the latter has tended to promote extensification and lower input-output production, whereas in Ontario the emphasis is on more sensible management of the existing ('industrial') farming system. Some common measures between Ontario and the European Union (EU) can be found, but essentially the EFP encourages a reduction in negative environmental outcomes arising from productivist agriculture. There is no suggestion of substituting a lower input-output system or of developing organic production. Indeed, participating farmers tended to view the EFP as complementary to their existing farming system but adding sensible precautionary measures designed to represent good stewardship of the land. An additional encouragement to participate was the availability of a small amount of financial support for taking 'environmental actions'. In some cases regulatory cross-compliance has been introduced so that farmers who are not implementing an EFP workplan are not eligible to receive income from other environmental schemes such as a Clean Water Project and a Healthy Futures scheme.

The author's interviews with key informants and a cross-section of Ontarian farmers (both participants and non-participants) revealed that most participating farmers tended to focus only upon certain aspects of their farm's operation rather than applying an environmental plan to the entire farm. So, in formulating individual EFPs, farmers evaluated only those activities associated with readily identifiable environmental problems. This means that, in common with several agri-environment schemes operated in the EU, the Ontario EFP is not applied on a whole farm basis. Individual farmers can omit or overlook certain environmental issues, either because they are deemed to be unimportant or they do not wish to pursue particular environmental actions. However, this individual interpretation represents a distinctive contrast to many European agri-environmental schemes where greater control 'from above' is exerted both in establishing the scheme's parameters and then in plan formulation and implementation on individual farms. In departing from the notion of implementing a predetermined package of measures and, instead, utilising a highly farm-specific blueprint, the EFP has been referred to as being 'needs focused' rather than 'solution focused' (Smithers and Furman, 2003). Ideally, the Ontario approach means that the measures adopted are tailored primarily to specific environmental circumstances and the goals of each farm and farmer. However, as there have been only relatively limited assessments of the environmental impacts of the EFP it is difficult to determine just what effects the scheme is having on wildlife habitats and biodiversity. The overall contribution to the sustainability of the farming system and of wider ecosystems also remains in question.

Despite concerns about the extent of the EFP's impacts on environment, it has been deemed to be successful by policy-makers and largely too by the farming community. As a result its application has been extended throughout Canada, first in the Atlantic provinces from 1996 and more recently in British Columbia (2003) and Alberta (2004). Starting in 2003 the Canadian government has allocated Cn$100 million over a four-year period to ensure that environmental farm planning will be delivered across the country. At the start of the programme it was estimated that EFPs were being followed by 8 per cent of all farmers in Canada (AAFC, 2004). In response to the new national Agricultural Policy Framework (APF) and the availability of ongoing federal support for environmental farm planning, a new phase of Ontario's EFP scheme was agreed in April 2005, with CN$57 million allocated by federal government and Cn$20 million from the provincial government through the Nutrient Management Financial Assistance Program (NMFAP) (OSCIA, 2005). A third edition of worksheets has been issued, with increased financial inducement for farmers who implement an individual EFP. This takes the form of applications for funds to the Canada-Ontario Farm Stewardship Program (COFSP), which will share costs with farmers (either 30 per cent or 50 per cent of eligible costs of implementation up to Cn$30,000) for specific beneficial management projects (BMPs) covering 25 designated categories relating to schemes designed to reduce environmental stress on the land (AAFC, 2005).

Conclusions

The continuing vision for ACAP is that the ACAP Boards should bring together and co-ordinate the activities of various disparate organisations to develop solutions to 'environmental' problems and thereby, as a by-product, help to create a more sustainable local community. In this respect it has similarities with the Local Agenda 21 (LA21) programme established after the Rio de Janeiro 'Earth Summit' in 1992. In particular, ACAP shares with LA21 a model of citizenship linked closely to the state and also the concept of multi-sector partnerships "within which members of local public, private and voluntary sectors have a role in contributing to the formation of a vision of local sustainable development and of then implementing that vision" (Whitehead, 2007: 204). In addition, though, the ACAP program also promotes an ecosystem approach to achieving community objectives. The use of natural systems boundaries is an important ACAP concept and allows for explicit ecosystem objectives with measurable indicators, such as water quality and air quality, percentage of solid waste recycled and sustained beneficial resource uses. The ecosystem approach also recognises the important linkages between economic and social well-being and the health of the environment.

It is undoubtedly the case that through the various projects implemented through the ACAP scheme, an environmental dimension is becoming a factor in previously unrelated areas, especially within the business community and certain aspects of economic development. Moreover, public consultation and participation in local decision-making has been extended and enhanced. This echoes findings for investigations of the impact of LA21 plans in Sweden (Jörby, 2002). As with ACAP, new opportunities emerged for individuals to engage in environmentally beneficial activities, and greater co-ordination of activities between different organisations was apparent. In Sweden direct environmental benefits were observed, as has been the case in the various ACAP areas. However, limitations to the Swedish LA21 plans were also noted, some of which find echoes in the ACAP scheme. In particular, a significant weakness in Sweden has been the heavy dependence on local government support and a top-down emphasis in which local people are only involved in an intermittent and consultative way. This has restricted the engagement of citizens in the process and so, ultimately, the LA21 projects will not be sustainable. In contrast, despite the input of federal and provincial funding and the important role of Environment Canada, there is evidence of ongoing citizen participation and involvement in such a way that there is a stronger bottom-up dimension than in the Swedish LA21. Moreover, ACAP has involved over 800 individual community action, science, knowledge sharing and capacity building projects to date (Lumb and Héalie, 2006).

The partnerships developed through the ACAP scheme represent a degree of 'citizen power' that elevates the assessment of public participation above basic levels recognised in formulations of a participatory continuum from a total lack of citizen involvement to total citizen control (Arnstein, 1969; Robinson, 1997b). The growth of a growing 'bottom up' culture within the ACAP groups gives real hope for future development of greater sustainability within these groups, and lends weight to the notion that sustainable development must necessarily include a strong citizen participation dimension alongside the traditional tripartite of the environmental (ecological), the social and the economic.

In terms of Ontario's EFP, the other case study examined in this chapter, one mark of 'success' is that this model is being transferred throughout Canada. Therefore this is ensuring that the 'needs focused' approach is being pursued within Canadian agriculture with respect to delivering certain environmental gains. As with the ACAP programme, responsibility for delivery has been partly removed from government and translocated to individuals or 'bottom up' organisations – in this case, farmers' representatives, with an emphasis on reducing farm-based pollution and delivering various other environmental benefits from productivist agriculture. However, this is largely retaining the industrial model of agriculture, partially tempered by marginal environmental gains. This might be regarded as a first step towards greater sustainability but at present it is not representative of a substantial move towards the generation of sustainable agriculture. In particular, it needs to involve a greater proportion of the farming community and it needs to be extended beyond delivery of a relatively narrowly defined set of environmental gains.

Overall, the two case studies illustrate that progress in the direction of greater sustainability is occurring within various sectors of the rural environment, but that these gains are uneven and are constrained by the nature of the economic system within which they are embedded. Clearly discernible environmental gains are apparent (especially in the ACAP scheme), but these are also heavily dependent upon the nature of the regulatory systems in operation and the evolution of 'ownership' of decision-making by the wider community. The nature of the linkages between government, business and societal stakeholders is crucial in determining how new ideas relating to sustainability *as a process* are received and then implemented. Where linkages are well-developed there can be growth of a consensus that recognises environmental gains are not necessarily at the expense of a thriving economy, and that such gains can generate their own economic and social benefits.

Acknowledgements

The author gratefully acknowledges support from the Department and Foreign Affairs and International Trade, Canada, who funded the two research projects upon which this chapter is based. Additional assistance from Environment Canada, the Ontario Soil and Crop Improvement Association, Mount Allison University and the University of Guelph is also acknowledged.

References

Agriculture and Agri-Food Canada (AAFC) (2004) Environmental farm planning coming to British Columbia. News release transmitted by CCN Matthews, 26.2.04.

Bidgood, M. (1994) *A study of actions by Simcoe County Environmental Farm Plan participants.* Guelph: Ontario Soil and Crop Improvement Association.

Caldwell, W. (2001) A municipal perspective on risk management and agriculture. *Great Lakes Geographer*, 8: 31–40.

Castelnuovo, R. (1999) Farm A* Syst/Home *A* Syst: a framework for voluntary action that is both effective and replicable. *Water Science and Technology*, 39: 315–322.

Dech, S. (2002) *ACAP's Science Linkages Initiative: a sound investment in science and community, 1997–2002.* Dartmouth, Nova Scotia: Environment Canada.

Environment Canada (1994) *Ecosystem Initiatives in Environment Canada: a synopsis.* Ottawa: Environment Canada.

Environment Canada (1995) *Guiding principles for ecosystem initiatives.* Ottawa: Environment Canada.

Environment Canada (2002) *An evaluation of the Atlantic Coastal Action Program: economic impact and return on investment.* Halifax, Nova Scotia: Gardner Pinfold Consulting Economists.

Environment Canada (2004) *Atlantic Coastal Action Program (ACAP): ACAP Phase III* http://atlantic-web1.ns.ec.gc.ca/community/acap/default. asp?lang=En&n=386354EA-1.

Fitzgibbon, J., Plummer, R. and Summers, R. (2004) *Environmental Farm Plan indicator survey.* Guelph: OSCIA for the OFEC.

Friedmann, J. and Weaver, C. (1979) *Territory and function: the evolution of regional planning.* London: Edward Arnold.

Green, D.G. (1987) *The new right: the counter-revolution in political, economic and social thought.* Brighton: Wheatsheaf.

Henderson, H. (1998) Noxious neighbours. *Planning*, 64(11): 4–9.

Jörby, S.A. (2002) Local Agenda 21 in four Swedish municipalities: a toll towards sustainability? *Journal of Environmental Planning and Management*, 45: 219–244.

Klupfel, E.J. (1998) Developing an understanding of and recommendations for promotion of the Ontario Environmental Farm Plan. Unpublished MSc thesis, Department of Rural Extension Studies, University of Guelph, Guelph.

Lumb, A. and Héalie, R. (2006) Canada's ecosystem initiatives. *Environmental Monitoring and Assessment*, 113: 1–3.

McNeil, T., Rousseau, F. and Hildebrand, L. (2006) Community-Based environmental management in Atlantic Canada: the impacts and spheres of influence of the Atlantic Coastal Action Program. *Environmental Monitoring and Assessment*, 113: 367–383.

O'Connor, Hon. D.R. (2002) *Report of the Walkerton inquiry: the events of May 2000 and related issues.* Toronto: Ontario Ministry of the Attorney General.

Ontario Soil and Crop Improvement Association (OSCIA) (2005) Next generation of Environmental Farm Plans launched in Ontario. http://www.ontariosoilcrop.org/ EFP/MediaRelease_NewProgramAnnounced.htm Accessed October 19, 2005.

Redclift, M.R. (1992) The meaning of sustainable development. *Geoforum*, 23: 395–403.

Robinson, G.M. (1997a) Community-based planning: Canada's Atlantic Coastal Action Program (ACAP). *Geographical Journal*, 163(1): 25–37.

Robinson, G.M. (1997b) Environment and community: Canada's Atlantic Coastal Action Program (ACAP). *London Journal of Canadian Studies*, 13: 121–137.

Robinson, G.M. (2006a) Ontario's Environmental Farm Plan: evaluation and research agenda. *Geoforum*, 37: 859–873.

Robinson, G.M. (2006b) Canada's environmental farm plans: transatlantic perspectives on agri-environmental schemes. *Geographical Journal*, 172(3): 206–218.

Sharpe, A. and Sullivan, D. (2004) *The effectiveness of the Annapolis River Guardians: Influences on public policy, decision-making and environmental awareness.* Annapolis Royal, Nova Scotia: Clean Annapolis River Project.

Shortall, S. (1994) The Irish rural development paradigm – an exploratory analysis. *Economic and Social Review*, 25: 233–60.

Smithers, J. and Furman, M. (2003) Environmental farm planning in Ontario: exploring participation and the endurance of change. *Land Use Policy*, 20: 343–356.

Stohr, W.B. and Taylor, D.R.F. (1981) *Development from above or below? The dialectics of regional planning in Developing Countries*. Chichester and New York: John Wiley and Sons.

Whitehead, M. (2007) *Spaces of sustainability: geographical perspectives on the sustainable society*. London and New York: Routledge.

Wilkinson, K.P. (1991) *The community in rural America*. New York: Greenwood Press.

Index

208 *Sustainable Rural Systems*